# CLASSROOM
# 2061

## Activity-Based Assessments in Science Integrated with Mathematics and Language Arts

Elizabeth Hammerman

Diann Musial

IRI SkyLight

TRAINING AND PUBLISHING, INC.

Arlington Heights, Illinois

**Classroom 2061: Activity-Based Assessments in Science Integrated with Mathematics and Language Arts**

Published by IRI/SkyLight Training and Publishing, Inc.
2626 S. Clearbrook Dr.
Arlington Heights, Illinois 60005-5310
800-348-4474, 847-290-6600
FAX 847-290-6609
info@iriskylight.com
http://www.iriskylight.com

Creative Director: Robin Fogarty
Managing Editor: Julia Noblitt
Editor: Amy Wolgemuth
Proofreader: Troy Slocum
Graphic Designer: Bruce Leckie
Illustration and Cover Designer: David Stockman
Formatter: Heidi Ray
Production Supervisor: Bob Crump

LCCCN 95-79277
ISBN 1-57517-004-3

1617C-3-97V
Item number 1347
06 05 04 03 02 01 00 99 98 97   15 14 13 12 11 10 9 8 7 6 5 4 3

# Dedication

To Alton—our mentor who gave inspiration and united two different worlds of thought.

To Donald—our colleague who continually provides guidance and support for our efforts.

A note of gratitude to Susan Klipp, a teacher who truly understands how students think and learn, for her creative contributions.

# Contents

## PART I
## THEORY

## PART II
# APPLICATION

Contents

# Foreword

New definitions and structures for science curriculum and new teaching strategies are difficult to attain. However, change in assessment strategies has been even more difficult. Teachers often return to quizzes, unit tests, and standard achievement measures as the only valid and objective indicators of learning. School administrators, boards of education, and the general public are also guilty of assuming that the only valid assessment strategies are the typical recall type of instruments and tests. New curriculum and teaching strategies are often assessed in terms of their effect upon standard measures of student learning.

For his doctoral dissertation, E. E. Zehr studied more than two hundred Iowa K–12 teachers who were actively involved with science-technology-society approaches and with the National Science Teachers Association and the Science Standards and Curriculum project (The University of Iowa 1991). He found that although teachers were willing to assess students in multiple domains, they used traditional test items to determine grades. Many attempted to write application and performance items; however, careful analysis revealed that nearly all of the assessment items resembled typical recall questions that focused on definitions, ideas, and skills taught directly in the classroom. These teachers seldom saw the importance of altering their assessment strategies.

Elizabeth Hammerman and Diann Musial have recognized the problem with assessment reforms. They write and speak to the importance of assessment in meeting new goals, matching new curriculum structures, and corresponding to the improvement in teaching practices. *Classroom 2061* provides a fine rationale for changing assessment to relate to the other facets of current reform efforts. The authors' philosophy, recommendations, and examples correspond to the ideas advanced in the National Research Council's *National Science Education Standards*.

Hammerman and Musial have developed their book in collaboration with some of the most creative upper elementary and middle school teachers in Illinois. The result is an easily understood and helpful book that encourages teachers to develop, try, and use performance measures in assessing their teaching and the learning of their students.

Performance assessments are defined as a set of tasks that include hands-on activities, criterion-referenced test items, and open-ended writing prompts. Through the multidimensional tasks, students are given various opportunities to show what they know and can do. The tasks assess content, skills, and habits of mind as well as students' ability to relate science to the real world.

It is important that students, parents, administrators, and others are involved in identifying desirable and appropriate performances. They should all be partners in encouraging learning and determining that real learning has occurred. Real learning is usually best indicated by the transfer of concepts and skills to completely new situations—evidence that students can be stimulated to provide routinely.

Assessment is a basic part of science. In fact, science cannot exist without it. Scientific inquiry begins with a question followed by a personal explanation (hypothesis), which in turn must be tested for its validity. This testing is assessment; it is the evidence that must be shared and accepted by others for a hypothesis about the natural world to become scientific knowledge.

Engaging students' minds in the learning process is the goal. In the February 1994 issue of *Educational Leadership* (How to Engage Students in Learning, 11–13), Vito Perrone offers a list of strategies for student engagement:

1.  Students have time to wonder and find a particular direction that interests them.
2.  Topics have a "strange" quality—something common seen in a new way, evoking a lingering question.
3.  Teachers permit—even encourage—different forms of expression and respect students' views.
4.  Teachers are passionate about their work. The richest activities are those "invented" by teachers and their students.
5.  Students create original and public products; they gain some form of "expertness."
6.  Students do something—e.g., participate in political action, write a letter to the editor, work with the homeless.
7.  Students sense that the results of their work are not predetermined or fully predictable.

Hammerman and Musial provide ideas for performance assessment that promote student engagement. They offer great ideas for developing rubrics to determine the extent and the degree to which real learning has occurred.

I recommend *Classroom 2061* to all teachers who want to learn how to make their assessment of learning more authentic. The ideas will help promote current reform efforts and provide great hope that these efforts will be more successful and longer lasting than all our past attempts at educational reform.

ROBERT E. YAGER, PROFESSOR
SCIENCE EDUCATION
THE UNIVERSITY OF IOWA

# Introduction

This book was written in response to many requests from schools to provide a clearly articulated set of performance assessments in science. Steeped in the sweeping ideas of Project 2061 as presented in the American Association for the Advancement of Science's *Science for All Americans* and *Benchmarks for Science Literacy*, we have formulated a set of carefully crafted performance assessment prototypes for the education practitioner.

To provide opportunities for students to show what they do and do not yet know and what they can and cannot yet do, multiple and varied assessment methods must be employed to sufficiently explore student learning. At the heart of these methods is the notion of a performance. Through performances, teachers can employ a variety of methods for students to demonstrate what they have learned. Assessment tools include learning logs, portfolios, peer interviews, teacher-student interviews, observations made by teachers and pupils, problem situations that are solved both individually and in a group, independent oral and written reports, group reports, the creation of products, and the development of videotapes. Educators are being asked to employ a wide variety of these tools in order to create student profiles that are based on valid and reliable assessments.

We have designed a set of authentic performance assessments that offer students the opportunity to demonstrate knowledge and multiple science abilities through communication, problem solving, inventiveness, persistence, and curiosity. The specific knowledge and science abilities are drawn from those identified in *Benchmarks for Science Literacy* and other national standards for science (NRC 1995) and mathematics (NCTM 1991). They are models for school districts and teachers to use in developing their own meaningful performance assessments. As prototypes, they illustrate a new way of looking at assessment in science, mathematics, and language. As assessments, they enable students to demonstrate their understanding of concepts, skills, and attitudes embedded within a single performance.

The performance assessments in this book do not deal with only one small component of the national goals; that is, they are not meant to measure one process skill at a time using some isolated task or activity. Rather, these assessments are rich, filled with opportunities for students to show various skills, concept understandings, and habits of mind. The performances allow students to examine the world in a variety of ways while simultaneously yielding reliable performance data.

## Assessing Understanding

Humans strive to make sense of the world around them; it is part of human nature to seek understanding. It is not surprising that both parents and students assume that schools are the vehicles through which understanding is achieved. But, what is understanding? This word is used in many different ways and contexts. If understanding is the desired outcome of instruction, educators must recognize the multidimensional aspects of the concept and carefully assess the congruence of their instructional practices to these different dimensions. For instance, in the context of the science disciplines, knowledge has three dimensions that relate to three levels of understanding. Each of these dimensions is characterized by an increasingly more sophisticated level of knowing.

### Knowledge as Information (Level One)

The first level of understanding includes a set of facts or information that one might accumulate and store internally as mental images. Informational knowledge is important merely as a first step toward reaching understanding. Traditional tests tend to focus on this level of knowledge for several reasons: l) paper-and-pencil questions dealing with factual information are readily available; 2) testing time to engage in expository questions is not needed; 3) more topics can be assessed at a lower level; and 4) a set of facts is often all the teacher knows or has time to teach about a topic. More sophisticated levels of understanding are seldom assessed and, therefore, left to chance.

### Knowledge as Relationship (Level Two)

This more advanced level of knowledge goes beyond the information or content that is part of one's mental image. It focuses on the relationships between mental constructs and experiences. At this level of understanding, students are able to generate justifications and explanations for their mental images. For instance, when a student not only identifies rocks by their names and observable properties but knows the rocks in terms of their mineral compositions and/or can describe the manner in which they were formed, the student has achieved a higher level of knowledge. At this level, students are able to relate rocks to other curricular content and processes and to the real world, thus demonstrating a broader and deeper understanding. Ultimately, the student views rocks as part of larger constructs such as matter, cycle, and change. This level of understanding can be effectively assessed in an activity-driven performance.

### Knowledge as Inquiry (Level Three)

The third level of knowledge goes beyond what one already knows *and* the connections to related concepts and phenomena. It challenges the boundaries of the learner's mental images and experiences. Through inquiry, students are able to think about their own thinking, critique what they know, and generate new questions that reach beyond the knowledge they already possess. Such questions are generated by encountering new experiences in a somewhat "risky" manner, since challenging the limits of what one knows includes admitting that one does not have all the answers or knowledge. For example, in a physics class, students might study various types of forces and see them in action. Students may learn the concept of force through a knowledge base (level one) and through firsthand experience at an amusement park (level two). The realization of force becomes a set of mental images coupled with experiential learning and problem solving. Once the student reaches a relational level of concept understanding within the realm of physics, the student can further investigate a concept through inquiry into the psychological and social forces operating in other contexts.

## Components of Meaningful Performance Assessments

The prototypic performance assessments in this book have been designed to go beyond the level of informational knowledge. Each assessment provides opportunities for students to explore their understanding of science, mathematics, and language through relationship and inquiry. The performance assessments permit students to explore and develop their own tests, ask questions, and articulate new relationships. The assessments enable students to exhibit at least two levels of understanding and allow teachers the opportunity to go beyond what is suggested in these activities. To accommodate such important educational goals, each assessment has been carefully constructed to include the following components or criteria.

### Thoughtful, Engaging Approaches

Insofar as possible, the performances promote higher-order thinking and help develop more complex cognitive functions. Students have an opportunity to display different levels of understanding during these performances: share ideas, generate new questions, create operational definitions, identify variables, and perform other tasks requiring complexity of thought.

### Rich Opportunities for Problem Solving

The prototypic performance assessments allow students to solve problems in a variety of creative ways. They encourage students to generate new applications, explore different problem-solving paths, and demonstrate understanding in a variety of ways. As such, these performances will be open-ended rather than closed, active rather than passive, authentic rather than contrived, and essential rather than tangential.

## Science Concepts

The context of the performances are centered on the basic concepts of science at the third through ninth grade level. In addition to the "big ideas" of science (e.g., matter, cycle, energy), the subtopics of science (e.g., magnets, water cycle, heat energy) provide the context for the performance tasks.

## Process Skills

Performances include demonstrations of the process skills of science, such as observing, classifying, measuring, making inferences, predicting, data collecting, and drawing conclusions. The prototypic performances assess a variety of process skills within each task.

## Habits of Mind

Throughout history, people have passed down values, attitudes, and perspectives about knowledge from one generation to another. These values, attitudes, and perspectives can be thought of as habits of mind because they relate to an outlook toward knowledge and learning and ways of thinking and acting. Habits of mind in the context of science include a willingness to take risks and the ability to formulate questions, exhibit curiosity, show persistence, analyze information with an open mind, and display a healthy skepticism. The prototypic performances require students to practice these habits of mind.

## Science-Technology-Society Connections

The prototypic performances are set within a real-life context. Students are asked to investigate science problems and questions embedded in the society in which they live. The performances also require students to consider the influence of science principles and technology on their world.

# The Performance Assessments in This Book

The prototypic performance assessments that comprise chapters 5 through 13 provide a variety of ways to assess students' concept understanding, process skill acquisition, habits of mind, and ability to make real-world connections. Most of the performances include more than one activity, a writing prompt, and a set of criterion-referenced questions that can be used in conjunction with the activities in the performance. Scoring the writing prompts and determining an acceptable score for the criterion-referenced questions has been left to teacher discretion.

Although the assessments in this book have a strong science component, the tasks have been constructed to reflect the new visions for mathematics and language arts as well. The tasks include many of the components of mathematical power, as discussed in chapter 2, and the discourse that so closely ties to language arts. The scoring rubric that is provided for each performance assessment attempts to identify the specific components and dimensions being assessed. (The scoring rubric is described in detail in chapter 3.)

Since dimensions often overlap across the disciplines, it is left to the teacher to decide whether measurement, for example, will be a mathematics or a science assessment. The decision is entirely arbitrary, since many dimensions innately belong to more than one discipline. The same is true for language arts. Communication is a dimension of both science and mathematics. Thus, the communication of results can be assessed in the context of mathematics, science, or language arts.

Chapter 4 offers a step-by-step analysis of each component of a prototypic performance assessment. This chapter is intended to guide you in using and developing more authentic assessments for your students.

# PART I

# THEORY

# WHAT IS SCIENCE?

# WHAT IS

The discipline of science is often thought of as a body of knowledge dealing with various aspects of the physical and biological world. Science is commonly referred to as biology, physics, earth science, and chemistry. Each of these subjects comprises a massive amount of information. The perception of science and, thus, science education as primarily concept understanding implies that the result of learning science is the acquisition of a knowledge base of facts.

More contemporary definitions of science and science education view concept acquisition as but one piece of a broader, richer picture. *Science for All Americans* defines science education as education in natural and social sciences, mathematics, and technology. This larger vision of science focuses on inquiry about the physical world in the context of society. This vision can be thought of as a puzzle that comprises an intricate interplay of several pieces: concept understanding, process skills, habits of mind, and science-technology-society connections. The pieces of science, just as the pieces in a puzzle, are not meaningful in and of themselves. Rather, each piece must be studied within the framework of the other pieces.

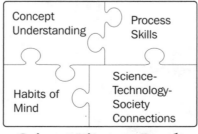

**Science Literacy Puzzle**

## Science Literacy

It is not enough for education to be self-fulfilling; science education needs to prepare citizens to deal with global, national, and local problems, such as population growth, loss of resources, and the effects of pollution,

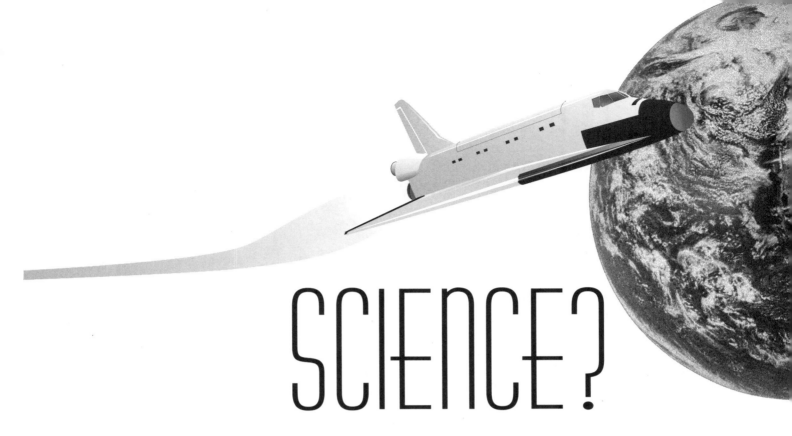

# SCIENCE?

disease, and social strife. The definition of the scientifically literate citizen includes the following characteristics (AAAS 1989):

- familiarization with the natural world
- awareness of how mathematics, science, and technology are related to one another
- understanding of key concepts and principles of science
- ability to think scientifically
- knowledge that science, mathematics, and technology are human enterprises and what that implies about their strengths and limitations
- ability to use scientific knowledge and ways of thinking for personal and social purposes

If acquiring these characteristics is deemed a worthy educational goal, then the key components of science literacy should be carefully and creatively integrated into the school curriculum. Such curriculum components derived from the list of characteristics above might include the following:

- a knowledge base emphasizing basic concepts and principles as well as more abstract concepts such as unity and diversity
- an understanding of the interrelatedness of science, technology, and society (S-T-S connections)
- strengths and limitations of science as a human enterprise
- ability to think and trust the thinking process and to develop problem-solving skills and processes

△ *The key components of science literacy should be carefully and creatively integrated into the school curriculum.*

Take a moment to reflect on the science curriculum you know and/or work with. Analyze it with regard to the areas noted above: concept understanding, interrelatedness, human enterprise, and thinking/problem solving.

The results of your analysis will assist you in determining the degree to which your educational program matches the new vision for science. Such an analysis is a crucial step in the development of meaningful assessments.

## Concept Understanding

The traditional fields of science are biology, physics, chemistry, and earth science. Environmental science might be considered a separate category or it might be included in earth science. Although it appears that these titles are universally understood, since textbooks are generally organized under these headings, the arbitrary designations do little to define the concept knowledge base of science. For example, many branches of science, such as oceanography and ecology, have biological and physical aspects as interrelated components of study.

A more appropriate portrayal of the concept areas of science is as a set of universally accepted "big ideas" that thread throughout major areas of study of the natural world. When viewed in this larger, broader context, a set of cognitive "frames" are identified that become the foundation for building and structuring knowledge over time (Musial and Hammerman 1993).

Because teachers often present science curriculum in the form of topics, it is necessary that they realize that most commonly taught topics are merely subcategories of bigger ideas. An example of a big idea of science is a cycle. A cycle can be defined as a continuous series of events that repeats itself over time. The events in a cycle can be repeated at regular or variable intervals. At the primary school level, the cycle of day and night may be introduced. The regularity of the day and night cycle can be easily observed in a short period of time. The cycle of the seasons takes longer to observe, but over the course of the year students can recognize the varying characteristics of seasonal change. Some cycles, like the water cycle, do not occur at regular intervals. Students may observe simulations of this cycle in a shorter time frame than what occurs in nature. Some life cycles can be observed in a relatively short time span, such as the life cycle of the darkling beetle, but other cycles, such as the rock-soil cycle or certain chemical cycles, are not as regular or as easily observed firsthand.

Regardless of which examples of cycles are studied, students need to understand that all the examples represent the larger construct of cycle and that they all have characteristics that allow them to fit this big idea. A big idea, or cognitive frame, that is introduced early in a child's life is known only through a few examples. Over time that cognitive frame expands and develops through exposure to multiple examples, experiences, and connections. Every cognitive frame has a broad meaning that the student will use to make sense of a more abstract, complex world.

△ *A more appropriate portrayal of the content areas of science is as a set of universally accepted "big ideas" that thread throughout major areas of study of the natural world.*

---

**"Big Ideas" of Science**

**force**          *time &*
                   *space*
**systems**

          **symmetry**
cycles      **& patterns**

---

The common characteristics between the cycle of the seasons and the life cycle of the frog will not be apparent to most students. In processing for meaning, teachers need to help students identify the examples as subcategories of big ideas. Whether they are dealing with a single activity or a larger unit of study, teachers need to help students identify similarities that exist between the subcategories (e.g., seasonal cycle, life cycle) and the big idea (e.g., cycle). Big ideas can also be useful because they cross disciplines. For example, cycles are found in history, mathematics, economics, and other areas of study.

The commonalities that exist within the knowledge base of science (as well as the other disciplines) provide a natural way to sort and classify information within the brain. Information relating to science becomes far more manageable when students realize that most of it will fit into a set of approximately twenty-five big ideas (Cox 1987). The state of Illinois, for example, has included the mastery of seventeen such concepts as a subgoal of one of its four state goals for science. Some of these big ideas are symmetry and patterns, matter and energy, cause-and-effect relationships, systems, organisms, relationship of structure to function, force, perception, time and space, and equilibrium. These big ideas, or concepts, are embedded in the many topics of science. For example, when studying butterflies, the important science concepts are cycle, interaction, and organism. These concepts need to be made explicit for students within the science unit about butterflies.

## Process Skills

The process skills of science (which also include a number of skills that are important in other disciplines) are a set of thinking strategies that scientists use in their activities. Beginning with the basic skills of observation, classification, and prediction, and moving to the more complex skills of hypothesizing, controlling variables, and drawing conclusions, process skills are the catalyst for thinking in science. Process skills are best used in hands-on activities where students engage in a problem-solving or investigative task while manipulating materials. The activity approach to learning science allows students to become active learners, using all of their senses and, to a degree, involving themselves emotionally; this enables students to make more meaningful connections within the brain.

Process skills are found in the procedures of an activity. The names of the skills should be used and reinforced by teachers as students use and develop these skills over time. These names will become familiar to students, enabling them to use the correct vocabulary when they explain their involvement in an activity. Although there is not a universal list of scientific process skills, the list in Figure 1.1 is an example of what is found in the literature.

---

## Scientific Process Skills

**Classifying:** arranging or distributing objects, events, or information and representing them in classes according to a method or system

**Communicating:** conveying meaning and feelings through oral and written words, symbols, diagrams, maps, or graphs

**Creating models:** displaying information by means of graphic illustrations or other multisensory representations

**Defining operationally:** naming or defining objects, events, or phenomena on the basis of their functions and characteristics

**Drawing conclusions:** making summary statements that follow logically from data collected during an investigation

**Formulating hypotheses:** making statements that are tentative and testable

**Identifying variables:** recognizing the characteristics of objects or factors in events that are likely to change under certain conditions

**Inferring:** giving a reasonable explanation for an observation

**Interpreting data:** analyzing data that have been obtained and organized by determining patterns or relationships in the data

**Measuring:** making quantitative observations by comparing to a conventional or nonconventional standard

**Observing:** becoming aware of an object or event by using any of the senses to identify properties

**Predicting:** making a forecast of future events or conditions

**Recording data:** collecting information about objects and events

**Using numbers:** applying mathematical rules or formulae to calculate quantities or determine relationships from measurement

**Verifying:** conducting additional tests to confirm or determine the adequateness of prior conclusions

---

**Figure 1.1**

# Habits of Mind

Another important component of science is a set of attitudes and dispositions, referred to as habits of mind, which relate to how a student views knowledge and learning. Scientists learn about the universe through direct involvement in laboratory and/or field investigations or experiments. Behaviors and ways of thinking that exemplify the attitudes and beliefs of working scientists can be practiced by students through inquiry-/activity-based science instruction. Dispositions such as curiosity, honesty, integrity, open-mindedness, respect for life, willingness to suspend judgment, and respect for data are some of the habits of mind students can develop through an inquiry-/activity-based program.

Through laboratory and/or field investigations, students can also practice safety, accuracy, good experimental technique, systematization of data, persistence, effective communication, analysis of strategies and results, and replication of the work of others. When students are allowed to perform the work of a scientist as part of the curricular program, they have the best chance of developing habits of mind that are highly regarded by those in science and by society as a whole.

# Science-Technology-Society Connections

Understanding the relationship of science to technology and society is also a critical component of science literacy. Science does not exist in a vacuum. Science is the world around us; it exists everywhere. Interest in science seems to wane when science is not seen as a vital component of our existence. As humans, we are a part of the natural world; to understand nature is to better understand ourselves.

In our current technological world, technology and society are intertwined with science in ways that are both intriguing and masterful. To study science without the awareness of its connections to the world deprives students of the context to which they, as members of society, are closely allied. Students are not always aware of how the concepts of science look in society. They do not fully understand the ways that politics, values, economics, and other social structures interact with the findings of scientists. When students recognize the connections of science to technology and society, they come to appreciate the value of scientific discoveries, and they become aware of the strengths and limitations of science in relation to social problems.

Most students are aware of some areas of science in the world, such as the exploration of outer space and the study of extinct dinosaurs. Through media and community resources, students are exposed to scientific fact as well as fiction. If all areas of science could be as real to students as the two examples cited, there would be a much greater interest in science not only during the relatively short span of the K–12 school experience but throughout life.

△ *When students recognize the connections of science to technology and society, they come to appreciate the value of scientific discoveries, and they become aware of the strengths and limitations of science in relation to social problems.*

## Methods of Teaching Science

One major factor influencing student interest in and respect for science is the method by which it is taught in schools. Children have a natural curiosity for the natural world and exhibit this interest in many ways outside of the school environment. Within the school environment, science instruction often consists of reading, writing, and memorizing facts. When learning science is presented as an opportunity to explore, investigate, research, and discover, it is exciting to students and fits naturally with their learning processes. The goals of science education can be met only through a strong commitment to an inquiry-/activity-based science program as the means for learning key concepts, developing skills, practicing scientific habits of mind, and making real-world connections.

## CHAPTER 2

# NEW VISIONS FOR MATHEMATICS AND LANGUAGE ARTS

*y*

*x*

*z*

# NEW VISIONS
# FOR

The disciplines of mathematics and language arts are going through a revolution. Both have experienced a paradigm shift. This shift includes a move away from the mastery of information to the attainment of a broad, interconnected knowledge base. Both disciplines are being restructured toward this new vision of knowledge where the emphasis is on providing instruction that enables the learner to analyze and communicate effectively in an ever-changing environment. To create authentic assessments in these disciplines, the new visions for mathematics and language arts must be carefully examined. Otherwise, the assessments will focus on the wrong components and, ultimately, provide data relevant to the old paradigms.

## The Vision for Mathematics

Perhaps more than any other discipline, mathematics has evolved into a massive information base filled with centuries of insights. Somewhere along its development, mathematics began to look like an encyclopedia of information about relationships among numbers and space. Schools began to amass this information into thicker and thicker textbooks filled with problems that required one right answer. Not surprisingly, teachers began to teach for these right answers, and students began to view mathematics as the mastery of problems that someone else developed. The problems became more and more abstract, and their relationship to the real world increasingly remote.

Recent efforts by the National Council of Teachers of Mathematics (NCTM) and the National Research Council (NRC) have led mathematicians, educators, and the general public to reconsider the essence and vision of mathematics.

# MATHEMATICS AND LANGUAGE ARTS

This effort has resulted in the development of two terms: *mathematical literacy* and *mathematical power*. Mathematical literacy includes having an appreciation of the value and beauty of mathematics and being able and inclined to appraise and use quantitative information. Mathematical power is the ability to do purposeful and worthwhile work, such as exploring, conjecturing, and using logical reason, as well as the ability to use a variety of mathematical methods effectively to solve nonroutine problems and possession of the self-confidence and disposition to do so (NCTM 1989).

## Mathematical Literacy and Mathematical Power

The terms mathematical literacy and mathematical power represent a new vision for mathematics, a vision that places mathematical reasoning, problem solving, communication, and connections at the center of mathematics. This vision attempts to de-emphasize convergent thinking and replace such thinking with divergent thinking. To do so requires a concomitant de-emphasis on mathematical problems with single answers to interdisciplinary problems with several thinking paths that lead to a variety of reasonable solutions. Computational algorithms, the manipulation of expressions, and paper-and-pencil drills must no longer dominate mathematics instruction. Beyond the standard fare of number concepts and operations, the school curriculum must include serious exploration of geometry, measurement, statistics, probability, algebra, and functions. Students should encounter, develop, and use mathematical ideas and skills in the context of genuine problems and situations. In doing so, they should develop the ability to use a variety of resources and tools, such as calculators, computers, and concrete, pictorial, and metaphorical models. Students must know and be able to choose appropriate methods of computation, including estimation, mental calculation, and the use of technology, and be able to engage in conjecture and argument as they explore and solve problems.

The new vision for mathematics is similar to the new vision for science. As these two disciplines attempt to recover their essences in the midst of a sophisticated technological society, they begin to focus on similar components: understanding, thinking or process skills, and dispositions or habits of mind. An examination of these components reveals the remarkable relationship between science and mathematics. Assessments in science that focus on these basic components will be potential assessments in mathematics as well.

△ *The new vision for mathematics is similar to the new vision for science. As these two disciplines attempt to recover their essences in the midst of a sophisticated technological society, they begin to focus on similar components: understanding, thinking or process skills, and dispositions or habits of mind.*

## Mathematics Understanding—Big Ideas

The first component of the new vision for mathematics relates to concept understanding, where the concepts are comprised of big ideas. This component of mathematics requires teachers to focus on large mathematical ideas. The ideas are not to be taught in isolation; rather, the NCTM standards emphasize that students should use these mathematical concepts (also called strands) to make connections among other concepts and across disciplines. Examples of these concepts include change, symmetry, equality, inequality, numeration, dimensions, ratio, perimeter, area, and synthetic and coordinate representations.

## Mathematical Thinking—Process Skills

The second component of the new vision for mathematics emphasizes thinking- or process-skill development. It focuses on the use of knowledge and understanding to analyze, conjecture, design, evaluate, generalize, investigate, model, predict, transform, or verify. This component is almost identical to that of science process skills. In fact, the process skills for science can be found in the list of mathematical skills generated by both the National Council of Teachers of Mathematics and the National Research Council.

This emphasis on the development of thinking has significant implications for assessment. Teachers cannot assess thinking skills without providing problems that require the use of these skills. Although memorization is a useful skill, it is not the focus of the new vision for mathematics. Performance assessments that require skills such as analysis, conjecture, speculation, and reflection will evoke student thinking more effectively than assessments comprised of paper-and-pencil exercises.

## Mathematics Tools and Techniques

Given the technological complexity of today's problems, the new vision for mathematics includes an emphasis on efficiently and effectively solving problems.

This vision requires teachers to assist students in the use of manipulatives. Using calculators and computers is no longer a tangential component of mathematics; it is an essential part of a student's mathematics education. Problems that students encounter in the classroom should be complex, but students should no longer be rewarded for solving such complexities the "long way." Rather, students should be encouraged to consider what tools can help to solve the problems most efficiently in a given situation.

The new vision for mathematics also encompasses the use of diagrams, tables, charts, and other concrete materials to assist in solving problems. The use of number symbols is not always the most effective way to present a solution. Graphs, drawings, and written or verbal presentations are often more effective than numeric expressions.

Such a changed vision has profound implications for assessment. Performance assessments need to focus on problems that require data-collecting and the use of organizing structures such as charts, diagrams, and tables. Manipulatives should be available to students as needed but should not be used indiscriminately. Students should be encouraged to select the correct tool(s) based on the demands of the problem. To use the calculator for every problem is inappropriate, as is solving all problems through paper-and-pencil or mental computation.

△ *Using calculators and computers is no longer a tangential component of mathematics; it is an essential part of a student's mathematics education.*

## Communication Skills

The final component of the new vision for mathematics focuses on communication. This component requires students to communicate results with various audiences for a variety of purposes. To fully assess mathematics without also assessing language skills is virtually impossible. As a result, the new vision for mathematics requires interdisciplinary assessments. To adequately assess learning in mathematics, teachers must assess language (specifically communication skills) simultaneously.

## The Vision for Language Arts

Language arts, like mathematics, has suffered because of its highly developed knowledge base. Throughout the last century, language arts instruction has grown to include huge amounts of information about decoding, phonics, vocabulary, reading, sentence structure, writing, literature, poetry, and more. Over time, language arts has been specialized and compartmentalized into smaller subdisciplines. Reading has been separated from writing, and literature from the development of language. However, efforts by the National Reading Teachers Association have bolstered support for the "whole language" movement. As the word "whole" implies, this movement (or new vision) encourages teachers and students to view language as an interrelated set of activities that focuses on meaning-making and effective communication. Within this vision, reading cannot be taught separately from writing, nor can language arts be taught separately from literature. The whole-language movement has transformed the way textbooks are written and the way literature is presented.

Whole language encourages teachers to approach language arts instruction as a three-fold process: meaning-making, comprehending the meaning of others, and communicating meaning to others. This three-fold process requires students to actively reflect on the meanings that they are always in the process of developing. Students who are taught to reflect on the meanings that they create tend to develop meanings that are more accurate, comprehensible, and useful to society. Whole-language instruction assists students in this reflective process by providing a variety of symbols for expressing and clarifying meaning. For example, students can draw, create graphs, sing songs, write poetry, and even use their bodies to express concepts that sometimes cannot be expressed in narrative form.

Whole language also encourages students to actively develop their ability to understand the meanings of others. Such understanding will occur if students are taught to use and identify many different types of symbols. Sensitivity to different ways of communicating helps students comprehend the meanings of others. Assessments that are true to this vision of language are interdisciplinary. Each time a student is asked to communicate, no matter what the discipline, that act of communication is part of whole language. Mathematics, science, and language arts are inextricably connected in the world of whole language.

△ *Students who are taught to reflect on the meanings that they create tend to develop meanings that are more accurate, comprehensible, and useful to society.*

## Implications for Assessment in Mathematics and Language Arts

To accommodate the new visions for mathematics and language arts, assessments must be reconceptualized. Mathematics assessments usually include a set of paper-and-pencil problems. Language arts assessments often focus on mechanics, also using paper-and-pencil tests. To reflect a reformed vision for mathematics and language arts, assessments need to be reformulated into significant tasks. These tasks provide the intellectual contexts for students' mathematical, scientific, social, and language development. They need to provide the stimulus for students to think in real-world contexts. Tasks that require students to reason and to communicate mathematically, scientifically, and socially are more likely to improve students' ability to make connections and promote greater understanding. Teachers need to select activities that create opportunities for students to develop cognitive knowledge, skill competence, increased interest, and positive dispositions. Activities conducted in science classes can also focus on mathematics and language arts, thus providing greater opportunities for developing interdisciplinary understandings.

Assessments that capture the new visions for mathematics and language arts need to be open-ended enough to permit discourse. Discourse refers to the multiple ways of representing, thinking, talking, and agreeing and disagreeing that teachers and students use to engage in the tasks. Discourse permits thinking to unfold and also affirms the importance of interactive, collaborative learning. Brainstorming, discussing, sharing results, and respectful questioning are skills that need to be practiced and sharpened. These skills, especially, relate to the communication aspect of mathematics and science; they also permit teachers to assess student understanding. Assessments that are rooted in activities that permit discourse will more readily lend themselves to the new visions for mathematics and language arts.

## CHAPTER 3

# DEVELOPING SCORING RUBRICS FOR

# PERFORMANCE ASSESSMENTS

# DEVELOPING SCORING RUBRICS

## FOR

A s has been shown in the previous chapters, the boundaries that have traditionally separated the disciplines of science, mathematics, and language arts are being challenged. Science now focuses on a set of big ideas, as does mathematics. The ideas look different in the contexts of the different disciplines, but they are essentially the same: symmetry, balance, equilibrium, cycles, and so on. The new visions require teachers to reconsider the concepts they currently teach and strive to teach concepts, or big ideas, that transfer across disciplines.

The same is true for process-skill instruction. It is now apparent that the thinking skills required in science are similar to those used in mathematics. Classification in mathematics is similar to classification in science. Completing an analysis in mathematics is like completing an analysis in science. This is also true for the scientific habits of mind. In mathematics these habits of mind are called mathematical dispositions. No matter what the term, these habits of mind tend to be similar throughout the disciplines. For example, the ability to ask clear questions is essential in mathematics, science, and language arts. Noting the similarities among the concepts, skills, and habits of mind of the different disciplines is very helpful when developing scoring rubrics, as it will assist in identifying the important features of an interdisciplinary task.

## Rubrics

One of the greatest challenges in developing performance assessments is the ability to identify the critical features of a task. This can be likened to taking an x-ray. An x-ray enables a physician to "see" the underlying bones and muscles that support the human body. In a sense, this is the view a teacher

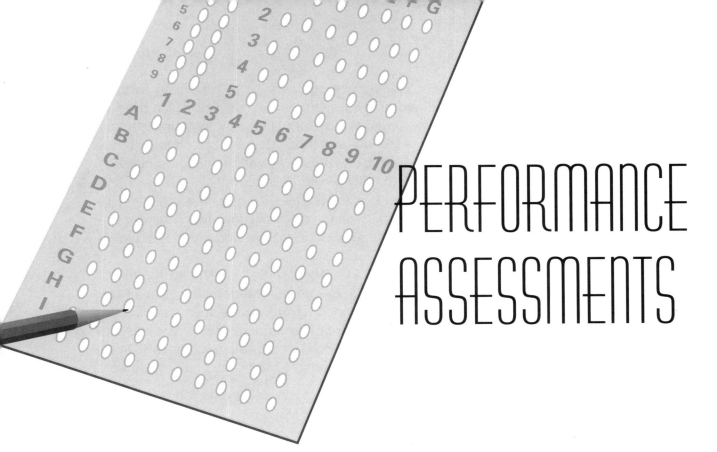

# PERFORMANCE ASSESSMENTS

needs when determining how to score or assess a performance. Every performance contains important underlying features or indicators of the critical dimensions of learning and inquiry. In science, these dimensions include concept understanding, process skills, habits of mind, and science-technology-society connections. Teachers must step back from the performance assessment and, like the physician's use of the x-ray procedure, look beyond the surface aspects of the performance to determine the most important dimensions. Identifying these indicators is at the heart of conceptualizing rubrics.

Rubrics can be thought of as a set of rules or criteria that provide direction in determining a score. In their simplest form, rubrics are nothing more than an answer key for a multiple-choice test. In this case, the only rule is to count the number of correct answers and, perhaps, cluster the answers into different subtest scores. The rubric or scoring key for multiple-choice, paper-and-pencil tests are simple; much more direction is required when developing rubrics for a rich performance assessment. Such performances can be assessed with holistic, generalized, or analytic scoring rubrics.

△ *One of the greatest challenges in developing performance assessments is the ability to identify the critical features of a task.*

## Holistic Scoring Rubrics

A holistic scoring rubric requires the teacher to think about or consider an entire task as a single entity or construct. Some holistic scoring experts maintain that holistic rubrics should not be developed prior to the implementation of a performance. According to holistic experts, the focus should be on the performance, not on predesigned rubrics. These experts encourage teachers to spend more effort developing rich performance opportunities for students than designing tasks to match predesigned rubrics.

To develop authentic holistic rubrics, teachers identify a worthy task, have students perform the task, and then view students' performances in their entirety. Once teachers have viewed the performances, they separate them into two groups: those that were adequate and those that were not. Once the performances have been separated, teachers review the adequate performances and select a subgroup that stood out as exceptional.

Then teachers review the performances that they determined were inadequate. They once again separate the inadequate pile into two subgroups: those performances that display serious inadequacies and those that are simply inadequate.

Finally, teachers return to each of the four groups, ranging from poor to exceptional, and identify (based on the evidence of student work) the characteristics that best describe each category or group. These descriptions constitute the final rubric and are used to score other students who complete the same performance. Performances that represent the salient characteristics of each of the four groups can be selected and used to better clarify the different categories of the rubric.

## Generalized Scoring Rubrics

Some experts contend that the disciplines are innately interconnected through similar dimensions and, therefore, that generalized rubric statements can be written for a variety of performances within the same discipline or across different disciplines. Some have begun to develop generalized scoring rubrics for dimensions that cut across the disciplines. They contend that scoring these underlying dimensions (concept understanding, process skills, habits of mind, and S-T-S connections) is best accomplished by viewing the student's entire performance. These experts argue that when teachers break up a performance into discrete pieces, or indicators, they may be distracted from appreciating the beauty and continuity of the entire task. Generalized scoring rubrics provide a framework for assessing a task without explicitly identifying discrete indicators.

A generalized scoring rubric requires the teacher to consider each dimension of a task as a single entity or construct. The rubric consists of categorical descriptions on a continuum. Teachers are required to select the category along the continuum that best represents the dimension as exhibited in the performance. An example of a generalized scoring rubric is provided in Figure 3.1.

A generalized scoring rubric can be converted to an analytic rubric by linking each dimension to a specific task. Once specific indicators are identified, an analytic scoring rubric emerges.

## Analytic Scoring Rubrics

The identification of individual, critical features inherent in an assessment task allows teachers to assess concept understanding, process skills, and

---

## Generalized Scoring Rubric Example

**Concept Understanding**

3   Demonstrates a complete understanding of important concepts or ideas
2   Displays an adequate understanding of important concepts or ideas
1   Displays an incomplete understanding of important concepts and ideas
0   Demonstrates serious misconceptions about concepts and ideas

**Process Skills**

3   Demonstrates mastery of important strategies and skills
2   Demonstrates important strategies and skills independently without significant error
1   Demonstrates important strategies and skills with guidance
0   Makes critical errors when carrying out important strategies and skills

**Communication Skills**

3   Consistently and effectively communicates by providing a clear main idea with support and elaboration
2   Communicates information by providing a clear main idea with little detail
1   Intermittently communicates a clear main idea
0   Rarely communicates a clear main idea

---

**Figure 3.1**

habits of mind as separate pieces. This discrete scoring system is called analytic scoring. For any task, the teacher decides which of the many features within the task will be assessed. Using the dimensions of concept understanding, process skills, habits of mind, and S-T-S connections as a guide, the teacher lists student behaviors or indicators that relate to any of these dimensions and assigns either a "yes" (there is evidence) or a "not yet" (evidence is lacking), or applies a binary scale of one or zero to each indicator. "Yes" (or one) means that the student exhibited the indicator; "no" (or zero) means that the student did not exhibit the indicator. For a simple experiment concerning temperature and plant growth, an analytic scoring rubric might look like the rubric in Figure 3.2.

Analytic scoring is simple and efficient. Such an approach not only clearly identifies the different dimensions of a task (concept understanding, process skills, habits of mind, S-T-S connections) but also clearly lists the specific features or indicators (measures, observes, predicts, describes, relates, works collaboratively) for each of the dimensions. These indicators can provide insight into a student's strengths and weaknesses and help diagnose problems with a student's understanding, skills, and/or behavior. For example, in the analytic scoring rubric in Figure 3.2, the student's ability to suggest a relationship between two variables (temperature and plant growth) might be assigned a "yes" or a "1." However, data collection over time may require more complex

---

## Analytic Scoring Rubric Example

**The student will**

suggest a reasonable relationship between
temperature and plant growth (process skill)      _____

record temperatures accurately (process skill)      _____

measure accurately in centimeters (process skill)      _____

record data for temperature (process skill)      _____

record data for plant growth (process skill)      _____

make a graph that represents the data (process skill)      _____

communicate results clearly (process skill)      _____

record data accurately (process skill)      _____

communicate honestly (habit of mind)      _____

draw a conclusion about the relationship (concept understanding/process skill)      _____

generalize to other plants and/or other conditions
(concept understanding/S-T-S connection)      _____

show persistence during the experiment (habit of mind)      _____

---

**Figure 3.2**

reasoning, and this indicator may be assigned two or even three points. The value of the points awarded for an indicator can be determined by teachers or students. Finally, the indicators can be clustered into subscores for each dimension; this permits each dimension to be represented by a separate total score.

Some dimension indicators are so complex that they may warrant additional consideration for scoring purposes. In the above example (fig. 3.2), when students generalize and apply their findings to other situations, they have an opportunity to elaborate on a concept. The teacher may wish to design a more elaborate scale in order to reward students for responses that show greater insight. For example, when students describe a relationship between two concepts, some descriptions might be more elaborate or provide more specificity than others. When an indicator implies a scale that is more complex than a simple "yes" or "not yet," the teacher is free to determine a greater range of scores.

## Rubrics for the Performance Assessments in This Book

We have determined that analytic scoring rubrics are most useful for clarifying the components of performances. Analytic scoring rubrics are user-friendly and enhance inter-rater reliability. For each performance in this book, we have identified the mathematics and science dimensions and the criteria for the writing prompts. The dimensions, and their corresponding indicators, can be used as they are or can be weighted depending on their importance or relative difficulty. Generalized and/or holistic rubrics can then be employed if needed.

# PUTTING
## IT ALL
# TOGETHER

## INTEGRATED

# PERFORMANCE

# ASSESSMENTS

# PUTTING IT ALL TOGETHER

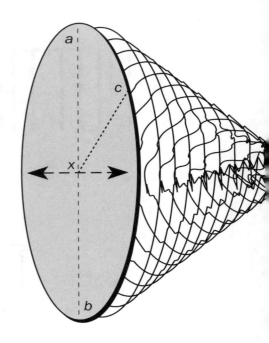

Although teachers have always been held accountable for what students learn in the classroom, the task of assessing student learning has often been relegated to reading from textbooks and recalling information. End-of-chapter tests are comprised of questions that can be answered by memorizing terms or passages from the text. Even when laboratory activities are included in the instructional process, the measure of student achievement is often tied to recalling factual information, identifying a correct response to a question, restating the proper definition, or using a correct mathematical formula. Assessment is generally measured in terms of what the student is able to show at the knowledge/information level; the acquisition of facts determines how informed students are about science.

Documents such as *Science for All Americans* call for reform in science education, emphasizing science's relationship to mathematics and technology and to the knowledge, skills, and habits of mind associated with all of these disciplines. National and state leaders emphasize the need to view science as more than a set of accumulated facts and theories. The recognition of science as multidimensional, having connections to mathematics and technology, requires administrators and teachers to redefine learning goals, revitalize instruction, and redesign assessments in ways that are different from traditional multiple-choice or forced-answer tests.

Assessment in this new mode is no longer limited to end-of-year or end-of-chapter, paper-and-pencil tests; nor is such assessment limited to memorized concepts. Traditional classroom assessments often fail to provide the opportunity for students to show their broad range of concept understanding, skill acquisition, habits of mind, and connections to other disciplines and

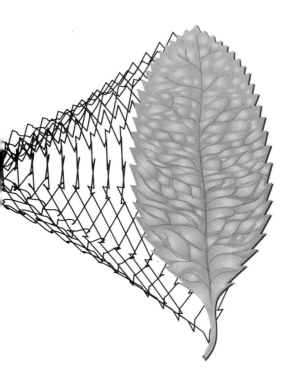

# INTEGRATED PERFORMANCE ASSESSMENTS

society at large. Meaningful assessments should be interesting and engaging in order to elicit information about what students know and can do.

## Assessment Strategies

Assessment plans that include a variety of strategies provide greater authenticity, that is, they provide multidimensional ways for students to communicate understanding. In this instance, more is better. Rich assessment plans might include such strategies as

- **Learning logs**—notebooks that contain written descriptions, drawings, data, charts, conclusions, inferences, generalizations, and other notes written by students
- **Portfolios**—a compilation of students' best work over time based on a variety of criteria
- **Interviews**—face-to-face discussions (student-to-student or student-to-teacher) about student learning
- **Observations**—recorded evidence of student learning and/or behaviors exhibited through classroom activities
- **Criterion-referenced tests**—multiple-choice questions clustered according to selected criteria allowing students several opportunities to display knowledge about each criterion
- **Writing prompts**—authentic situations or questions that require students to apply their understanding and communicate effectively through writing

- **Products**—integrated student-designed works that summarize student understandings
- **Activities**—experiences that require students to manipulate materials and reflect in meaningful ways

Performance assessments should be multifaceted tasks that focus on important dimensions of learning in a more comprehensive manner than any of the individual assessment strategies cited above. The performance assessments in this book comprise activities that link together a variety of assessment strategies into a meaningful whole. Such performance assessments provide a rich profile of student learning.

## Designing Integrated Performance Assessments

Meaningful performance assessments allow teachers to assess concept understanding, process skills, and habits of mind through a cohesive set of activities. These assessments should have the same qualities and characteristics of good instruction; they should involve students in active learning and be interesting and engaging.

As we work with science teachers to design meaningful, integrated performance assessments, we begin by asking them to consider what these performances would look like—what a student might "do" as part of, or following, a unit of instruction to show growth and development in concept understanding, skill acquisition, habits of mind, and connecting science to technology and society. Our experience shows that there is no magic formula for developing performance assessments.

*△ Meaningful performance assessments allow teachers to assess concept understanding, process skills, and habits of mind through a cohesive set of activities.*

A meaningful performance assessment is an activity or series of activities designed to gather information concerning what the student knows or can do. Ideally, the performance task should assess one or more levels of concept understanding and the ability to perform and/or apply process skills and habits of mind. The total assessment task must include one or more relevant activities and may include criterion-referenced test items, a writing prompt, and/or journal entries.

We have identified ten steps that we believe are critical to designing meaningful, integrated performance assessments. These steps are not necessarily sequential. Like juggling balls in the air, teachers need to keep all of the steps in motion while maintaining a unified perspective.

## 1. *Consider standards.*

Review national, state, and local documents to identify the important concept, skill, and habit of mind goals that are relevant to the age/grade of the students for whom the performance is being developed. Documents might include *Science for All Americans, Benchmarks for Science Literacy, National Science Standards,* and *Curriculum and Evaluation Standards for School Mathematics,* as well as documents that include state and local or district goals for learning.

## 2. Examine and discuss behaviors of scientists "in action."

How do scientists use concepts and skills in the work they do? What habits of mind, or dispositions, do they exhibit in their work?

Videotapes showing scientists in action are available from The New Explorers series (Public Media Education, 800-343-4312). You can view these videos or have informal discussions with working scientists and engineers from the community.

## 3. Relate components of science to your grade level.

Consider how components of 1) science concepts (including the applications and connections of concepts to the real world and other areas of the curriculum), 2) skills (including thinking skills and process skills), and 3) habits of mind that working scientists use compare with science at your grade level. Identify important concepts, skills, and habits of mind that are part of your science program and would be meaningful to assess.

## 4. Design a context for what you want to assess.

Think of a context for assessing some of the important components of science that you identified. The context may be an activity or a series of activities that relate to an investigation, a problem-based situation, a decision-making situation, the clarification of an issue, the development of a point of view, or the generation of a project, product, or invention.

The context should provide an opportunity for students to show any or all of the following related to science:

- knowledge and comprehension of concepts, application of concepts, and connection of concepts to real-world contexts
- ability to problem solve and exercise thinking skills
- ability to perform and/or apply process skills
- behaviors that exhibit habits of mind

Consider important components of other disciplines to assess as part of the performance task. For example, components of language arts might be assessed through an oral presentation of a science project. Such a performance might look like this:

> The students will demonstrate knowledge and use of basic science vocabulary when explaining a dinosaur diorama they made that includes the features of the habitat; the types and availability of food, water, and space for the animal; and the unique physical characteristics of the particular type of dinosaur they researched. They will use a variety of references and resources leading to the development of their habitat and animal models. Students will explain how they used researched evidence to make inferences concerning the extinction of their dinosaur. They will describe similarities and differences between extinct animals and animals that exist today.

## 5. Clarify and embellish the meaningful performance assessment.

As you design the performance task, build in as many opportunities as possible for students to show what they know and can do. Create a rough draft of the performance task, then go back and revisit the important concepts, skills, and habits of mind from science and other disciplines. Are there opportunities to enrich the task to include more of the important components? Besides activities, would other assessment tools, such as criterion-referenced items or writing prompts, be needed to get a more complete assessment?

△ **As you design the performance task, build in as many opportunities as possible for students to show what they know and can do.**

## 6. Identify important elements of the performance to assess.

Consider learning activities done prior to the performance assessment to be sure that students have had adequate preparation for the assessment task. In the example in step four, students might have studied the following: reptiles as a group of animals; dinosaur types; characteristics of herbivores, carnivores, and omnivores; factors (climatic, ecological, etc.) affecting habitats; habitat-animal relationships; food chains; animal structure and function; how to conduct science investigations; and/or how to research information through reading, museum visits, and interviews. Then, students would be assigned or would choose a dinosaur to study in detail.

Allow for input from students whenever possible when designing the performance. Identify the key elements of learning that relate to concept understanding, thinking and process skills, and habits of mind and that can be assessed through the performance and fit into the performance task. Connections to mathematics, technology, and society should also be included whenever possible. The set of identified elements establishes the performance criteria—the critical aspects or dimensions of concepts, skills, and habits of mind that students will have an opportunity to demonstrate through the performance. A list of key elements, or dimensions, that might be assessed in the example in step four are as follows:

### Concepts
Describes the dinosaur type and structure.

Explains the relationship between structure and function.

Shows and describes the habitat using the diorama.

Explains what evidence provided scientists with information about
    this species.

Describes the probable food chain that existed for this animal.

### Skills
Uses a variety of resources for information.

Classifies the animal type as herbivore, carnivore, or omnivore and
    tells why.

Makes a scale model of the dinosaur and describes the ratios used.

Infers why this animal became extinct.

Demonstrates a variety of skills used in an oral presentation.

### Habits of Mind

Communicates information honestly.

Demonstrates a respect for evidence.

Explains that what we know now about extinct species may change as scientists gather more evidence (open-mindedness).

### Science-Technology-Society (S-T-S) Connections

Explains how knowledge of dinosaurs has changed over time and infers why.

Provides names of resources available in the local community or elsewhere—places, people, multimedia, etc.—for learning more about dinosaurs or the field of paleontology.

You may decide not to assess all of the possible dimensions during the performance. It is important to determine which dimensions are most appropriate to assess at the time.

## 7. *Establish a rubric for assessment.*

Using the key elements of the task, list the learning dimensions and the indicators of the learning dimensions that you can assess through the performance. The learning dimensions are the concepts, skills, and habits of mind to be exhibited during the task. The indicators are what students will do in the task to show knowledge and/or understanding of concepts (e.g., identify and describe the components of the water cycle), application of skills (e.g., make a prediction; measure the mass of an object), and behaviors that exhibit habits of mind (e.g., communicate results of an investigation; show honesty and accuracy in recording data and information).

When creating a rubric for a performance, list the indicators in the order they are done in each activity. For example, one activity might have indicators listed in this order:

- Describes a plan for investigating the question.
- Predicts how one variable affects another.
- Measures the distance between points A and B.
- Records data.

By using a binary system (yes or no; 1 or 0), notation can be made next to the indicator as to whether or not the student showed evidence of learning. In some cases it may be appropriate to include a third category, as in "all information provided," "some information provided," and "no information provided." We recommend, however, that the scoring system include no more than three catagories.

If you wish to include subjective criteria based on things such as the aesthetic qualities of a project or product or the interest level of a topic selected by a student, these indicators can be listed at the end of the rubric along with the possible assessment categories. For example, for aesthetic qualities the categories might be "aesthetically pleasing," "aesthetically adequate," and "aesthetically inadequate."

By considering each indicator separately, you can document whether or not a student demonstrated concept understanding, skill acquisition, habits of mind, and S-T-S connections through the performance. The use of the evidence/no evidence system is analytic scoring. As explained in chapter 3, each aspect of learning is considered independently in analytic scoring. You may also develop a criterion or criteria for a quality performance and assign a point value for good, better, and best performances. A description of what each of the identified categories means is necessary in order to fairly assign credit for the quality of the performance.

*△ By considering each indicator separately, you can document whether or not a student demonstrated concept understanding, skill acquisition, habits of mind, and S-T-S connections through the performance.*

## 8. Determine what an acceptable performance looks like.

Teachers must determine the number of points that a student needs to earn in order to show that learning has occurred. This is done best by identifying what the categories of "exceeds," "meets," and "does not meet" expectations look like in assessing the performance task. Students may be able to assist in defining a quality performance. This step could also be called setting standards for an acceptable performance.

## 9. Communicating the performance assessment.

It is important to have a complete and accurate description of the performance assessment for your own understanding and for that of other teachers who might be using the assessment task. The following framework can be used to describe the performance assessment.

For the whole performance, provide the following:
- **Title of the Performance Assessment**
- **Rationale for the Performance**

  Provide a rationale for what you will assess and why it is important. You might discuss reasons why you selected certain concepts, skills, and habits of mind to assess. You might also include some beliefs you have about the teaching/learning process that determined your decisions. Considerations of the curriculum and/or the population of students might be included. The rationale should also include a brief description of the activity or activities as well as other assessment tools that make up the total assessment task.

- **Criterion-Referenced Test Items and Writing Prompt**
  Include these when appropriate to the activity.
- **Scoring Rubric**

For each activity, provide the following:

- **Title of Activity**
- **Materials**
  Provide a list of equipment and materials needed.
- **Description of Activity**
  Tell what students will do in the activity.
- **Presenting the Activity**
  Explain how to present the activity to students (what to say; how to say it) and provide management suggestions and other information that teachers need to know or that would help them to conduct the activity smoothly.
- **Student Response Sheet**
  Include all appropriate information for students to have when they are conducting the activity: title of activity; investigative question(s); clear directions for the activity; and spaces or lines for writing/drawing responses, data tables, graphs, and/or summary statements/ conclusions.

## 10. Field test the performance assessment and revise as needed.

Once you have developed a task that seems to have merit, pilot the performance with students. Ask students to evaluate the task based on how meaningful it was and whether or not it related to the learnings of the past year. Be sure to use the rubric you created to evaluate student projects and products from the performance. Often students will surprise you with their efforts, and you may find that the expectations you had when developing the rubric are different from those that students were able to meet. Use student comments and reactions to revise the activities of the performance task; use student work to clarify and revise the rubric.

## An Invitation

We have attempted to present a rationale for designing meaningful, integrated performance assessments that goes beyond mere verbal descriptions. In this book, we have provided examples or prototypes of what such performance assessments might look like in the classroom. Some of the examples are tied to specific topics in science; others simply provide a framework that allows teachers to select the context for the performance.

The major impetus for the performance assessments presented in this text comes from *Benchmarks for Science Literacy* by the American Association for the Advancement of Science. Specific benchmark statements from this book are identified for each of the performance assessments. These statements correspond to the concept understanding, skills, habits of mind, and real-world relationships a student might be able to show through the performances.

The performances are assessed through a variety of strategies: motivating activities, criterion-referenced tests, and writing prompts. The activities that comprise each performance are essential because they offer students multisensory experiences and opportunities to actively demonstrate their understandings and abilities in science. Criterion-referenced tests can be used to further probe students' understanding of concepts; and the writing prompts allow students to elaborate and apply what they have learned to a real-world context.

To meet national, state, and local standards of excellence, the prototypic performance assessments we offer in this book include the following:

- Inviting, engaging student-centered questions about the physical world and our relation to it
- Critical dimensions of science as described in *Science for All Americans* and *Benchmarks for Science Literacy*, including big ideas, process skills, and scientific habits of mind
- Rationales for the performances and ideas for ways to present activities
- Lesson plans for hands-on activities with materials lists, student response sheets, and other prompts to record student thinking
- Different types of questions requiring different types of thinking
- Criterion-referenced tests
- Writing prompts that connect science to society and technology
- Flexible analytic scoring rubrics that can be tailored to individual needs

△ *The activities that comprise each performance are essential because they offer students multisensory experiences and opportunities to actively demonstrate their understandings and abilities in science.*

An analytic scoring rubric is presented at the end of each performance assessment. Indicators are listed in the order in which they occur in the activities. Although many indicators are listed, teachers are encouraged to select those indicators that are of most importance to their unique settings. Teachers can also add other indicators. Each indicator is coded as follows:

C = concept understanding
S = process skills
H = habits of mind
STS = science-technology-society connections

Only habits of mind that are inherent in the activity are listed. Note that many other habits of mind, such as perseverance, honesty, and the ability to work in a collaborative group, may be assessed over time. The performance assessments may lend themselves to assessing a rich assortment of indicators, but teachers will need to add these as they fit their own teaching style and instructional techniques.

A key purpose of the writing prompt is to permit students to express their ideas in a more open and creative manner as they apply science to a real-world context. Responses to the writing prompts can indicate concept understanding and students' ability to apply the science concepts to a new situation. These writing prompts may also be a way of determining attitudes and values. Teachers are encouraged to determine their own communication skill indicators and to assess the dimensions of written and oral presentation skills if they wish.

In addition to a writing prompt, most performance assessments include a set of criterion-referenced, multiple-choice test items that correspond to the main concepts dealt with in the performance. Determining an acceptable total score for the multiple-choice questions has been left to teacher discretion. Note that the correct answer for all multiple-choice questions in this book is the letter "a." Teachers will need to vary the answers as questions are selected for classroom use.

We hope that these examples and "words of wisdom" assist teachers in designing performance assessments for their unique environments. To this end we, the authors, invite teachers to use these assessments, to edit them, and to make them an integral part of their diverse curricular contexts. No greater praise can match that of seeing these prototypic performances used and transformed through the creative talents of our fellow educators. We anticipate that this is but the beginning of an important revolution in the area of assessment and that the future will be filled with rich, meaningful, integrated performance assessments.

# PART II

# APPLICATION

## CHAPTER 5

# Adventuring Back in Time

## Targeted Process Skills

- observing
- classifying
- inferring
- collecting data
- comparing
- scale modeling
- measuring (length and mass)
- reasoning
- using coordinates

## Rationale

This performance assumes that students are familiar with the basic concepts relating to fossils and paleontology. They should know that fossils are the remains of animals and plants that have been preserved naturally. Students should be familiar with casts and molds, imprints, and mineral-replaced remains.

For activity 1, students will need to know how to make a scale drawing and be able to locate points on a grid by using the axis symbols that define the coordinates. If students have not had experience making scale drawings, you may want to conduct this part of the performance as a teacher-directed activity. In activity 2, students will be asked to measure in centimeters and to find the mass in grams. They will be asked to describe, compare, and infer, focusing on the characteristics that make it possible to identify a fossil and to distinguish it from others. Then, in activity 3, students will compare similarities and differences among the features of their fossils and those of animals that are alive today.

### PROJECT 2061 STATEMENTS

from *Benchmarks for Science Literacy*

Fossils can be compared to one another and to living organisms according to their similarities and differences. Some organisms that lived long ago are similar to existing organisms, but some are quite different (p. 123).

• • •

Similarities among organisms are found in internal anatomical features, which can be used to infer the degree of relatedness among organisms. In classifying organisms, biologists consider details of internal and external structures to be more important than behavior or general appearance (p. 104).

• • •

Scale drawings show shapes and compare locations of things very different in size (p. 223).

• • •

Make sketches to aid in explaining procedures or ideas (p. 296).

• • •

Find and describe locations on maps with rectangular and polar coordinates (p. 297).

PERFORMANCE
ACTIVITY

# Can You Dig It?

## Description of Activity

Students will be given a fossil-rich "site." They will be asked (or helped) to make a scale drawing one half the size of the site and label the coordinates on a grid. Using simple tools, students will excavate four different fossils from the site. They will describe the fossils and use coordinate symbols to indicate the location of the specimens on the scale drawing. Fossils may be presented in the pans so that the findings for each student will be identical. Or each student or student group can generate their own data by varying the fossil positions from pan to pan. Teachers may decide which situation is best for them.

## Presenting the Activity

Tell students they have been invited by the local museum paleontologist to assist with an exploratory dig at a fossil-rich quarry. Their task is to assist the paleontologist in identifying the site by drawing a map one half the size of the site. Students should mark off two centimeter sections along the top and the sides of the map. They can label the top of the grid with letters and the sides with numbers. Students will search for the fossils, describe them, and record the location where they were found on the map. (By varying the positions of the fossils for each student, students can each generate their own data.)

Figure 5.1 is an example of a site map that could be developed by students. Students may create many different variations of this type of map; however, each map should be approximately one half the size of their site.

### MATERIALS

❏ Large rectangular baking pans or cardboard trays filled with about two to three centimeters of sand or cat litter (cardboard trays from canned food may be obtained from grocery stores—approximately 26 cm x 36 cm x 5 cm)

❏ plastic spoons or other small tools to use to dig through the sand

❏ plastic fossils (or other real fossils, if possible) of crinoid, brachiopod, pelecypod, and petrified wood

❏ fossil kits from Creative Dimensions, P.O. Box 1393, Bellingham, WA 98227 (optional)

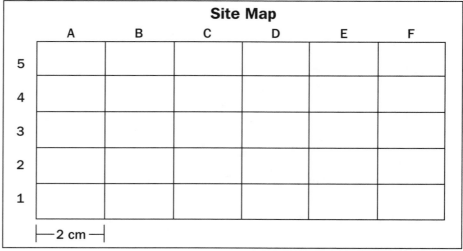

**Figure 5.1**

# Can You Dig It?

**How can we "map" the location of our fossil finds?**

After weeks of preparation with the paleontologist, you have arrived at the site for the "dig." It is an abandoned sedimentary rock quarry that, over time, has been covered with loose material. Buried in the material are fossils for which you will search. To begin you must record observations of the site, including a description and the size of the area to be investigated.

**Description** (material, color, landscape features, etc.)

_____

**Size of area** (in centimeters)

_____

Make a scale drawing of the area on a separate sheet of paper, one half the size of the actual area you are investigating. Mark off the drawing in 2 cm blocks so it looks like a grid. Label the grid with letters across the top or bottom and numbers along the side.

Be sure to draw the lines of the grid on your scale drawing so you can identify the location of your fossils when you find them. Fossil locations will be identified by letter and number. For example, C 2 would be the third square from the left or the right intersecting with the second one from the bottom or the top (depending on how you label your grid).

Using the digging tools provided, carefully probe the landscape until you find four fossils. Write a description of each fossil and give its coordinate location on the data table below.

| | **Description** | **Coordinates** |
|---|---|---|
| W | _____ | _____ |
| X | _____ | _____ |
| Y | _____ | _____ |
| Z | _____ | _____ |

Next put each letter W, X, Y, and Z or a drawing of each fossil on the grid to show the location where each was found. Your scale drawing becomes a map of your dig. How could your map be used by another explorer to find the location where your fossils were found?

_____

_____

_____

_____

_____

_____

_____

_____

_____

_____

_____

_____

_____

# Additional Lesson Ideas of Your Own

PERFORMANCE
ACTIVITY 2

# Now, Picture This

## Description of Activity

Students are asked to make a drawing of each of the four fossils. This is an important skill researchers need in order to record the intricate features of their finds. Students will measure the length, width, and mass of each fossil. Then, they will study each fossil's characteristics to determine if any resemble organisms that are alive today.

## Presenting the Activity

Discuss with students the fact that scientists record as much information as possible when doing research in the field. Making careful drawings is an important way of recording characteristics and unusual details of the fossils. Recording the size and the mass provides additional information. As an aid in identifying the fossils, it is important to note whether the fossil resembles anything that the student (or researcher) is already familiar with, such as characteristics of other known organisms.

**MATERIALS**

❏ fossils
❏ hand lenses
❏ drawing pencils
❏ metric rulers
❏ balances
❏ sets of weights

# Now, Picture This

**Do some of the fossils look similar to organisms we see today?**

The four specimens you excavated from your site are fossils of organisms that lived long ago. Draw each specimen in the space on the data table below. Measure the length and width (at the longest and widest points) of each fossil and record your findings on the table. Use the balance and weights to find the mass of each specimen in grams and record the results. Some of the fossils resemble currently existing organisms. Make inferences as to what organisms the fossils look like.

| Drawing of Specimen | Length (cm) | Width (cm) | Mass (g) | Organisms Fossil Resembles |
|---|---|---|---|---|
| | | | | |
| | | | | |
| | | | | |
| | | | | |

## PERFORMANCE ACTIVITY 3

### MATERIALS

- ❏ fossils
- ❏ hand lenses
- ❏ reference books for fossils
- ❏ fossil kits (optional)

# Here's Looking at You, Kid

## Description of Activity

Comparing similarities and differences among fossils and between fossils and other known organisms gives scientists clues to evolutionary patterns.

## Presenting the Activity

Students are often interested in knowing the names of fossils. Ask them to begin by comparing the fossils to one another. Do they have any characteristics in common? Are any two alike? Challenge students to find at least one way two of the fossils are alike.

Next give students access to field guides and reference books for fossils and let them find the type and name of each specimen. Have students record the names on their worksheets. From these resources, students can often find names or pictures of similar present-day organisms. Students should research and learn as much as they can about their four fossils and the organisms they resemble.

Name: _____

Date: _____

# Here's Looking at You, Kid

**How do fossils compare to one another and to other organisms?**

## Comparing Specimens to One Another

Use a hand lens to study your specimens. Compare them and list one way any two are alike.

_____

_____

## Comparing Specimens to Other Organisms

You may use reference books to help you find the name of each fossil and the names of any present day organisms that are similar to each fossil. Record the specimen names, the names of the organisms they resemble, and the characteristics of the fossils that are similar to the present-day organisms.

| | Specimen Name | Similar Organism | Similar Characteristics |
|---|---|---|---|
| W | _____ | _____ | _____ |
| X | _____ | _____ | _____ |
| Y | _____ | _____ | _____ |
| Z | _____ | _____ | _____ |

## Optional Task

What interesting information do scientists know about various fossils? (for use with Creative Dimensions Fossil Kit or another fossil kit)

For each of the fossils you found, locate the information card or research information about different fossils in your fossil kit. Use the information to find the following data for each fossil:

**W**

| Name | Age | Origin | Type of Fossil |
| --- | --- | --- | --- |
| _____ | _____ | _____ | _____ |

**Two Items of Interest**

_____

_____

**X**

| Name | Age | Origin | Type of Fossil |
| --- | --- | --- | --- |
| _____ | _____ | _____ | _____ |

**Two Items of Interest**

_____

_____

**Y**

| Name | Age | Origin | Type of Fossil |
| --- | --- | --- | --- |
| _____ | _____ | _____ | _____ |

**Two Items of Interest**

_____

_____

**Z**

| Name | Age | Origin | Type of Fossil |
| --- | --- | --- | --- |
| _____ | _____ | _____ | _____ |

**Two Items of Interest**

_____

_____

In what ways can information from the past help us understand the present?

_____

_____

_____

_____

_____

_____

_____

_____

_____

# Adventuring Back in Time

Note that the correct answer for all multiple-choice questions is letter "a." Teachers will need to vary the answers as questions are selected for use.

1. We can learn about the earth and organisms that lived millions of years ago by studying
   a. fossils
   b. sea shells
   c. living things
   d. igneous rock

2. Which of the elements below is NOT a type of fossil?
   a. sandstone
   b. casts
   c. molds
   d. petrified wood

3. Fossils allow us to study organisms that
   a. lived long ago
   b. lived recently
   c. are currently living
   d. never lived

4. Jane found a fossil that looked like a shell. She knew it was a fossil because it was made of
   a. minerals that replaced the shell
   b. the same thing as shells
   c. sand
   d. human-made materials

5. Fossils _____ look like organisms that exist today.
   a. sometimes
   b. always
   c. never

6. Scientists have learned a lot about how organisms evolved by studying
   a. fossils
   b. rock layers
   c. igneous rock
   d. present-day animals

7. When scientists find a fossil, they compare it to
   a. fossils and present day organisms
   b. fossils only
   c. present day organisms only
   d. nothing else

8. When scientists find a fossilized piece of a dinosaur, they try to reconstruct the whole animal. They do this by
   a. comparing the fossil to the structure of other dinosaurs
   b. studying the rock it came from
   c. studying living organisms
   d. comparing the fossil to present day dinosaurs

9. Plant and animals that exist today are
   a. similar to some fossils and different from some fossils
   b. similar to all fossils
   c. different from all fossils
   d. totally unlike fossils

# Making a Photographic Journal

Any expedition is a one time experience; the same events will not likely be repeated. Once fossils are removed from a site, their position, structure, and location can never be accurately recorded again. It is critical that precise records are kept of the dig.

On your expedition to discover fossils, you were asked by the paleontologists to act as the photo journalist in order to capture the unique characteristics of the trip. The scientists asked for four entries in the journal, each consisting of a snapshot and a written explanation of the image. Use your imagination to create unique features of your expedition. You are not limited by the actual site you experience; you may create a more interesting expedition in your photo journal.

Entries in the journal might include

- a picture and description of the site, complete with researchers and equipment

- paleontologists "in action" showing the processes they go through to take the fossils from the ground

- the finding of a fossil that is similar to a shell of an animal that might be found on a beach today

- pictures and descriptions of other fossils found at the site

- other unique and imaginative features of your experience

For each journal entry, provide a title, a picture, and an explanation of the picture using at least two sentences.

Be prepared to show your photo journal at a National Convention of Paleontologists. Have fun!

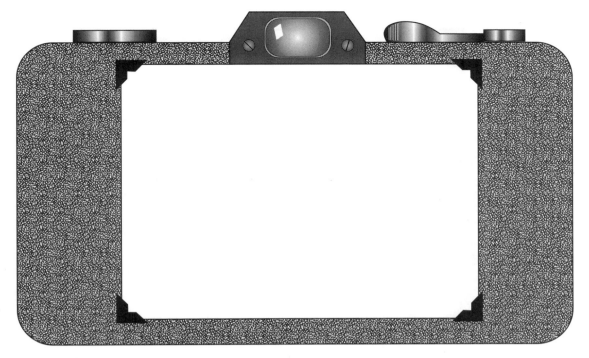

**Description of photo:** _____

_____

_____

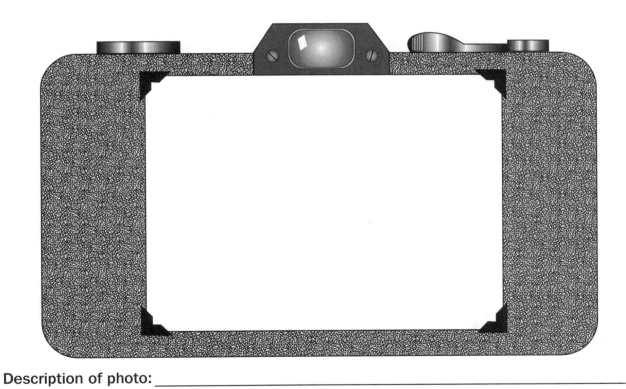

**Description of photo:** _____

_____

_____

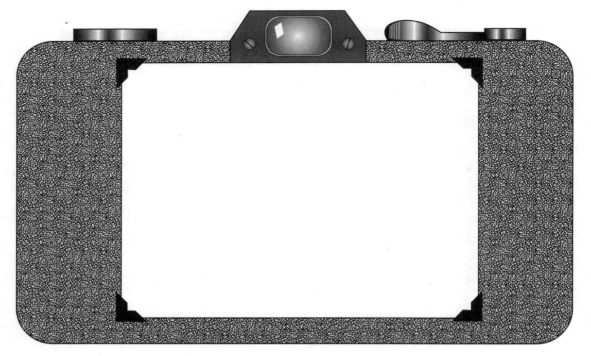

**Description of photo:** _____

_____

_____

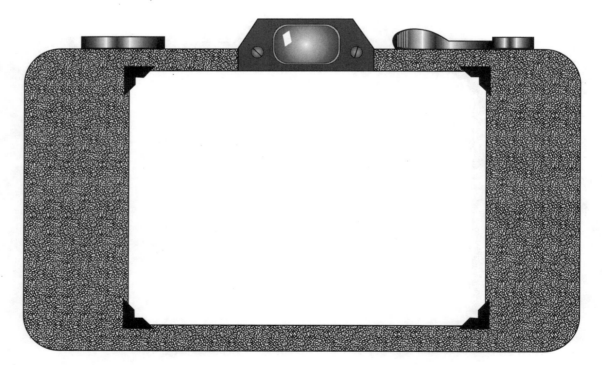

**Description of photo:** _____

_____

_____

*IRI/Skylight Training and Publishing, Inc.*

| Indicators | Dimensions |
|---|---|

## ACTIVITY 1 **Can You Dig It?**

| | |
|---|---|
| Writes a description of the site | S |
| Measures the size of the site in centimeters | S |
| Draws a model to scale | S |
| Labels the coordinates | S |
| Describes each fossil | S |
| Gives coordinates of locations | S |
| Puts W, X, Y, and Z correctly on the scale drawing | S |
| Gives a logical answer regarding the use of the map | C, S |

## ACTIVITY 2 **Now, Picture This**

| | |
|---|---|
| Draws the four specimens showing characteristics | S |
| Measures and records the length, width, and mass of each fossil | S |
| Makes inferences based on observations | S |

## ACTIVITY 3 **Here's Looking at You, Kid**

| | |
|---|---|
| Lists one way fossils are alike | C, S |
| Finds and records names of W, X, Y, and Z | C |
| Identifies organisms similar to W, X, Y, and Z | C, S |
| Lists characteristics that are similar to W, X, Y, and Z | S |

**Suggested Scoring Criteria for Writing Prompt:** Journal entries should include four pictures that are related to the student's experience in the activity. Each picture should have a clearly written description.

# The Dog Washing Business

## CHAPTER 6

## Targeted Process Skills

- observing
- measuring (volume and temperature)
- collecting data
- reasoning
- drawing conclusions
- problem solving
- graphing

## Rationale

This performance assumes that students have had experience reading thermometers in degrees Celsius and measuring the volume of liquids in milliliters. Students will learn about heat loss and change in temperatures in activity 1. In activity 2, students will explore changes in temperature over time, collect data, and make a graph to communicate their findings. They will also be asked to make a prediction by extending the line on their graph (extrapolation). In activity 3, students will use their new knowledge to develop an action plan for business, applying what they learned in activities 1 and 2 to a real-world situation.

---

### PROJECT 2061 STATEMENTS

from *Benchmarks for Science Literacy*

When warmer things are put with cooler ones, the warm ones lose heat and the cool ones gain it until they are all at the same temperature (p. 84).

• • •

Graphs can show a variety of possible relationships between two variables (p. 219).

• • •

Practical reasoning, such as diagnosing or troubleshooting almost anything, may require many-step, branching logic (p. 233).

---

# The Hot and Cold of Dog Washing

## Description of Activity

Students will measure the volume of water in milliliters and measure the temperature in degrees Celsius. (Fahrenheit equivalents are not necessary since all the relative information is given.) Students will discover what happens when cold water is mixed with hot water. Hot water should be at least 40°C (between 100–110°F); it may be necessary to heat the water using a hot plate. Be careful that the water is not too hot. Ice water or cold water from the tap may be used for the cold water. The more extreme the water temperatures, the more interesting the activity will be. In processing the results, students will determine if water gained or lost heat (energy transfer).

## Presenting the Activity

Tell students that any time a person wants to offer a service or start a business, there are many things to consider and investigate. In a dog washing business, the washers need to be concerned about the temperature of the water they use to wash the dogs. Since the available water is cold, they need to find out what happens when hot water is added to cold water. Students will use this information to help their friends Willie and Anita plan their business strategy.

### MATERIALS

❑ student thermometers measuring °C

❑ beakers with ml markings, or plain beakers or cups and a graduated cylinder with ml markings

# The Hot and Cold of

# Dog Washing

**What happens when hot water is mixed with cold water?**

Willie and Anita decided to start a dog washing and grooming business over the summer to earn money to go to camp. Since the water they had available was rather cold, they looked for ways to warm the water before washing the dogs. First, they decided to do some activities to learn more about the transfer of energy and the processes they might use to get the conditions they needed to start their business. They have asked you to help by doing the following test.

Measure 200 ml of cold water into one beaker (A) and 200 ml of hot water into a second beaker (B). Using a thermometer, measure the temperature of the water in each beaker in degrees Celsius. Record data below:

Beaker A _____ °C          Beaker B _____ °C

If the contents of beaker A were poured into beaker B, predict what the temperature of the water would be in degrees Celsius.

I think the temperature would be _____

because _____

_____

Now, pour the contents of beaker A into beaker B. Stir. Record the temperature of the water.

The temperature of the water is _____

How did your prediction compare to the actual temperature?

_____

_____

_____

Consider the water that was in beaker A. Did it lose or gain heat after it was mixed with the water in beaker B? _____

Consider the water that was in beaker B. Did it lose or gain heat after it was mixed with the water in beaker A? _____

What conclusion can you draw?

_____

_____

_____

Following the test, what advice can you give Willie and Anita that might help them plan their business?

_____

_____

_____

_____

_____

# Additional Lesson Ideas of Your Own

PERFORMANCE
ACTIVITY 2

## MATERIALS

❏ student thermometers measuring °C

❏ beakers of water (one marked A, one marked B)

❏ clock with minute hand

# Time and Temperature

## Description of Activity

Students will use the data they have collected about the temperature of the water in beakers A and B to answer the operational question: What will happen to the temperature of the water in thirty minutes? This activity will need to be done immediately following activity 1 since heat loss will occur right away, or activity 1 can be repeated as part of activity 2. Students should have experience watching time and collecting data using a thermometer. They can record the data and make a line graph to show the change in temperature over time.

A graph may be provided by the teacher, if appropriate, or students may be required to construct their own. Make sure the axes are labeled properly and the points are connected by a smooth line. Students will also make predictions about what the temperature will be after forty minutes based on the curve of the line that connects the points (extrapolation).

## Presenting the Activity

Tell students that Willie and Anita may have to let the water they are going to use sit for a while before they use it. Willie and Anita wonder what will happen to the temperature of the water over time. They ask, What will happen to the temperature of the water in thirty minutes? They can immediately take the temperature of the mixture of water in the beaker, and then they can take it again after ten minutes, twenty minutes, and thirty minutes.

Students should collect and then graph their data. A graph can be provided (or created) where temperature is plotted on the vertical axis and time is plotted on the horizontal axis (see fig. 6.1). Since the data show a loss of heat over time, the line connecting the points will be "downhill." Ask students to predict what the temperature will be after forty minutes, based on what they discovered in the investigation. Students should show their prediction on the graph by extending the line with a broken line and placing a point on the line at the forty-minute mark.

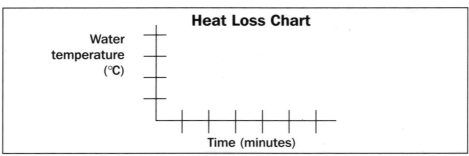

**Figure 6.1**

# Time and Temperature

## What will happen to the temperature of the water in thirty minutes?

You have found the temperature of the water when the contents of beaker A and beaker B were mixed. How will the temperature of the water change after ten minutes? twenty minutes? thirty minutes?

Predict what the temperature will be after thirty minutes.

I predict it will be _____

because _____

Take the temperature of the water after ten minutes, twenty minutes, and thirty minutes. Record the temperatures on the data table below.

| Time | Temperature | Loss of Heat |
|------|-------------|--------------|
| Present | _____ | |
| 10 minutes | _____ | #1_____ |
| 20 minutes | _____ | #2_____ |
| 30 minutes | _____ | #3_____ |

Predict what the temperature of the water will be after forty minutes.

    I predict it will be  _____

    because  _____

    _____

Now, make a temperature and time graph in the space provided below. Connect the three temperature points with a smooth line. Put a dot on the line at the forty-minute mark that corresponds to your prediction. Show an extension of the line to this point using a dashed line.

# Additional Lesson Ideas of Your Own

## PERFORMANCE ACTIVITY 3

### MATERIALS

❏ student thermometers measuring °C

❏ beakers with ml markings, or plain beakers or cups and a graduated cylinder with ml markings (same as activity 1)

# Setting Up Business

## Description of Activity

Students will be given some information and asked to help develop a reasonable plan for arriving at a usable water temperature. Since this is a problem-solving activity, results may vary. Some students may use what they discovered in activity 1, and some may not. Students may want to experiment with water using the beakers and thermometers, as they did in activity 1. Or they may prefer to make paper-and-pencil calculations. If the temperature of the cold water students have to work with is 20°C, it will allow for an easier mathematical solution to the problem.

## Presenting the Activity

Tell students that Anita and Willie need to develop a plan for getting the water to a comfortable temperature for washing the dogs. The tap water is too cold by itself. Their dad will provide hot water, but it is too hot for washing the dogs. The children must come up with a plan. Obviously, a plan for getting water to the proper temperature will require some thought and calculation. You may want to allow students to brainstorm criteria for a "good" plan. Each criterion identified by the children could be given a point. If students are not able to decide on the criteria for a good plan, the teacher should let students know what the scoring criteria will be. Criteria might include logic, organization, use of a thermometer at regular intervals, recording data, graphing the data, and drawing a conclusion. Throughout this discussion of a good plan, you may wish to refer to water gaining and losing heat as the "transfer of energy."

# Setting Up Business

Willie and Anita decided they needed water that was about 30°C to comfortably wash the dogs. The water they have to work with is only 10°C. This water needs to gain heat to be useful. Their dad told them he would provide hot water from the water heater in their home; it is 40°C. This water needs to lose heat to be useful.

Using the information you learned in activity 1, develop a plan for Willie and Anita in the space below so that they can take advantage of their dad's offer.

_____

_____

_____

_____

_____

_____

_____

_____

_____

_____

_____

_____

# The Dog Washing Business

Note that the correct answer for all multiple-choice questions is letter "a." Teachers will need to vary the answers as questions are selected for use.

1. When ice cream is taken from the freezer and left on the counter, it
   a. gains heat from the room
   b. loses cold to the room
   c. gets colder the longer it is out
   d. does not change temperature

2. The process of "thawing out" means the object must
   a. gain heat
   b. gain cold
   c. lose heat
   d. lose cold

3. José takes a mug of hot chocolate outside on a cold day. In a short time, he discovers that the liquid in the mug
   a. loses heat
   b. loses cold
   c. gains heat
   d. gains cold

4. Sometimes when it is cold, Mary takes a hot-water bottle to bed with her. In about two hours, she observes that the bottle has changed temperature. Most likely it
   a. lost heat
   b. gained heat
   c. lost cold
   d. gained cold

5. During the winter, heat is added to a room to warm it. If the heat were turned off, what would happen to the air in the room? It would
   a. lose heat
   b. gain cold
   c. lose cold
   d. gain heat

6. During the winter, Dan takes care of the animals in the barn. He adds a half bucket of warm water to the half bucket of cold water in the barn. The full bucket of water would then be
   a. warmer than it was
   b. colder than it was
   c. the same as it was

# Advertising Our Business

Willie and Anita have permission to distribute flyers advertising their business to the houses in their neighborhood. As an advertising and graphic-design specialist, you have been called to assist them in developing a flyer that would convince customers that their service is better than other dog washing businesses in the area.

Here are some questions and items you might consider for the flyer:

- What does Willie and Anita's service offer to dogs?
- Why would dogs like the service?
- How will they ensure that dogs get a comfortable bath?
- Quotes from dogs
- Information about Willie and Anita
- Pictures of the qualified staff in action
- Pictures of satisfied customers
- Competitive costs
- Other services and products available

The flyer should be designed on an 8.5 x 11 inch sheet of paper. It may be organized, folded, colored, etc., any way you wish.

| Indicators | Dimensions |
|---|---|

### ACTIVITY 1 **The Hot and Cold of Dog Washing**

| | |
|---|---|
| Measures 200 ml of water | S |
| Measures the temperatures of the hot and cold water | S |
| Predicts the temperature of the mixture | S |
| Measures the temperature of the mixture | S |
| Compares the prediction to the actual temperature | S |
| Describes what happened to the water in beaker A | S |
| Describes what happened to the water in beaker B | S |
| Draws a conclusion | S |
| Gives advice regarding findings | C, STS |

### ACTIVITY 2 **Time and Temperature**

| | |
|---|---|
| Predicts with logical explanation | S |
| Collects and records data | S |
| Records loss of heat accurately | S |
| Makes a second prediction with logical explanation | S |
| Constructs a line graph | S |
| Labels axes | S |
| Shows prediction on graph | S |

### ACTIVITY 3 **Setting Up Business**

| | |
|---|---|
| Solves the problem by developing a logical plan | S |

(Additional indicators and dimensions for the plan may be developed and scored accordingly.)

**Suggested Criteria for Writing Prompt:** The flyer should have information that describes the business, be the specified size, and be persuasive. Students may help determine the specific criteria for scoring. You may ask students to assess other students' flyers.

# May the Force Be with You

## CHAPTER 7

## Targeted Process Skills

- identifying properties
- observing
- collecting data
- inferring
- predicting
- designing tests and data tables
- communicating and applying findings
- using manipulatives
- measuring

## Rationale

This performance assumes that students are familiar with magnets and have done some preliminary work investigating magnetic properties with different objects. Students should know that metallic objects are attracted to magnets, as opposed to plastic, glass, or wood. In activity 1, students will learn that not all metal objects are magnetic and will make inferences about metal objects that exhibit magnetic properties. In activity 2, students will be asked to design a test to determine how far away objects can be from a magnet and still be attracted to it, in order to show their ability to measure and express an understanding of a magnetic field. In activity 3, students will set up a test to determine what happens when the poles of magnets come together, in order to show their understanding of the effect magnets have on one another. The performance ends with an opportunity for students to apply their understanding of forces within a field.

## PROJECT 2061 STATEMENTS

### from *Benchmarks for Science Literacy*

Without touching them, a magnet pulls on all things made of iron and either pushes or pulls on other magnets (p. 94).

• • •

People can often learn about things around them by just observing those things carefully, but sometimes they can learn more by doing something to the things and noting what happens (p. 10).

• • •

Offer reasons for findings and consider reasons suggested by others (p. 286).

• • •

Make sketches to aid in explaining procedures or ideas (p. 296).

• • •

When people care about what's being counted or measured, it is important for them to say what the units are (p. 292).

# Metals that Attract

## Description of Activity

Students are given a variety of metallic objects and a magnet. They will test each object to determine if it is attracted to the magnet. Students will record data by using an X to show which objects are attracted and will explain their findings.

## Presenting the Activity

Tell students they will be examining a variety of metallic objects using a magnet. Instruct them to read the directions and complete the data sheet by writing their responses in the appropriate spaces. It is important that inferences students make are based on the observations. Thus, if students observe that not all metallic objects are magnetic, they might infer that only certain metals are attracted to magnets. Students should realize that evidence, such as not all metals are magnetic, is what leads scientists to ask questions about the properties of materials. Their understanding of the usefulness of magnets in the real world is the focus of the open-ended question which helps students make an S-T-S connection.

## MATERIALS

- ❏ foil
- ❏ penny
- ❏ nickel
- ❏ paper clip
- ❏ brass fastener
- ❏ pin
- ❏ bottle cap
- ❏ bar, ring, or horseshoe magnets

# Metals that Attract

**Are all metal objects attracted to magnets?**

I think _____

because _____

Test the objects. Place an X next to the object(s) that are attracted by the magnet.

| Object | Attracted? (X = yes) |
|---|---|
| foil | _____ |
| penny | _____ |
| nickel | _____ |
| paper clip | _____ |
| brass fastener | _____ |
| pin | _____ |
| bottle cap | _____ |

What did you observe?

_____

_____

_____

Based on your observations, what inference(s) can you make about the objects that are attracted to magnets?

_____

_____

_____

_____

_____

_____

_____

_____

Think of an example of when a magnet could be helpful in accomplishing some task. Describe the task and explain how the magnet would help.

_____

_____

_____

_____

_____

_____

_____

_____

_____

_____

_____

_____

_____

# Additional Lesson Ideas of Your Own

PERFORMANCE
ACTIVITY 2

## MATERIALS

- ❏ bar, ring, or horseshoe magnets
- ❏ metal objects that are attracted to magnets from activity 1
- ❏ metric rulers

# Close Encounters

## Description of Activity

Students are asked to design an experiment to test an investigative question about whether objects must be in contact with the magnet in order to be affected by it. Students must make a prediction, describe the test and/or show it in a diagram, and perform the experiment. They are asked to design their own data tables and report the results. If students are not yet ready to design their own data tables, teachers may assist students in developing an appropriate data table or the class may decide on a single test and data table that all students will use.

## Presenting the Activity

Explain to students that they should follow the directions on the student data sheet and make a prediction with an explanation. Tell them that they will then conduct the test, record their data, draw a conclusion, and describe a magnetic field. Students should either design their own tests and create their own data tables or they should use the test and data table format designed by the class.

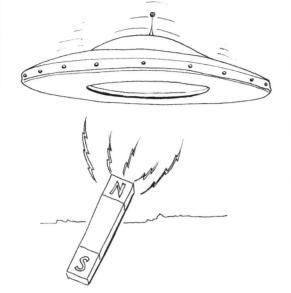

# Close
# Encounters

**How far away can an object be from a magnet and still be attracted to it?**

I think _____

because _____

Using the magnet and the metal objects, design a test that would answer the above question. Describe the test by writing or drawing it in the space provided.

Perform the test. Develop a data table to show your findings. Draw the table below. (If the distance you measure is less than 1 centimeter, use the symbol <. For example, for a measurement of approximately 1/2 centimeter write < 1 cm as a finding on the table.)

Write a conclusion.

_____

_____

_____

_____

_____

_____

_____

_____

_____

_____

_____

_____

Based on what you have observed, draw what a magnetic field might look like in the test you conducted.

PERFORMANCE ACTIVITY 3

## MATERIALS

- ❏ two bar or ring magnets per person or team
- ❏ rulers
- ❏ string
- ❏ scissors (If you are using a ring magnet, put a red sticker dot on the N pole and a blue sticker dot on the S pole.)

# Coming ATTRACTions

## Description of Activity

This activity investigates what happens when the poles of two magnets come together in the following combinations: N-N, N-S, S-S. Students should conclude that like poles repel while opposite poles attract. Students should also determine that objects do not have to be touching the source of the force acting on them in order for the force to exert a push or pull. The open-ended question at the end asks students to think about forces that affect people in their everyday lives. The "big idea" of force applies not only to magnetic force but to other forces, such as gravitational, buoyant, electrical, and centripetal. Students may also describe forces that are applied by humans in their examples.

## Presenting the Activity

Tell students that they will find out what effect magnets have on other magnets when the poles of the magnets come together. Instruct them to read the directions and complete the data table and questions. For the open-ended question, tell students to think of other forces that affect people or objects. You may relate the effect to other forces students have studied in science or social studies, but allow students to give the examples. Students may describe such forces as gravity or they may think about force in a social context as something that affects behavior, such as a force exerted by laws, police, or gangs.

# Coming ATTRACTions

**What happens when the poles of two magnets come together in the following combinations: N-N, N-S, S-S?**

Identify the north and south poles of a bar magnet (or red/blue sides of a ring magnet).

Place the north pole (red side) of one magnet near the north pole (red side) of the other magnet. Record your observations below.

_____

_____

_____

_____

Place the north pole (red side) of the first magnet near the south pole (blue side) of the second magnet. Record your observations below.

_____

_____

_____

_____

Finally, place the south pole (blue side) of the first magnet near the south pole (blue side) of the second magnet. Record your observations below.

_____

_____

_____

Summarize your findings.

**Magnetic poles**                 **Observations** (attract=X; repel=0)

North/North (red/red)                    _____

North/South (red/blue)                   _____

South/South (blue/blue)                  _____

What conclusion can you draw about the behavior of the magnets?

_____

_____

_____

_____

How is a magnet a type of force?

_____

_____

_____

_____


List two examples of forces in the environment. Describe each force and tell how each affects people or objects.

_____

_____

_____

_____

_____

_____

_____

_____

_____

_____

_____

_____

_____

_____

_____

_____

# May the Force Be with You

Note that the correct answer for all multiple-choice questions is letter "a." You will need to vary the answers as questions are selected for use.

1. Which of the following will be attracted by a magnet?
   a. paper clips
   b. plastic cubes
   c. rubber bands
   d. paper strips

2. When two magnetic poles are put close together the result is
   a. a push or pull
   b. no affect
   c. spinning of the magnets
   d. gravitational pull

3. A magnet produces a type of force because it
   a. pushes or pulls on objects
   b. pulls on glass objects
   c. has no effect on objects
   d. pushes on glass objects

4. An object that pushes or pulls on another produces a
   a. force
   b. chemical
   c. change
   d. concept

5. In order for a magnet to attract an object, it must be made of
   a. iron
   b. plastic
   c. rubber
   d. glass

6. Sally put a strong magnet on the table near a compass that shows direction. She observed that the needle on the compass moved. She inferred that the needle of the compass was made of
   a. iron
   b. plastic
   c. glass
   d. aluminum

7. A magnet has a north pole and a south pole. Two bar magnets are hung by strings so that the two north poles are next to one another.

   In this situation the magnets will
   a. push away from one another
   b. pull toward one another
   c. spin
   d. have no affect on one another

8. When Mark put the south pole of one magnet near the north pole of another magnet, he found that they
   a. attracted
   b. repelled
   c. pushed away
   d. had no effect

9. Which diagram shows the correct way to put two bar magnets away so that the two ends will attract?

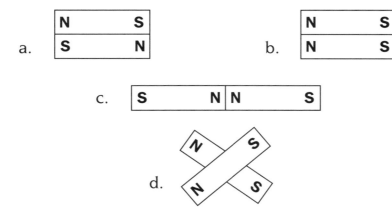

# A Friend in Need

Your cousin Terry who lives in another town is in charge of an aluminum recycling project. She has been given donations of all sorts of metal objects including some aluminum. She knows she has to separate the aluminum from the other metals, but does not have an effective way to do this.

Since you were instrumental in helping to design the first space robot that went to the moon, you have been asked to help your cousin design a robot to help sort the metal for the recycling project. Use what you have learned about magnets and metals in your design. Draw a picture of what the Metal Monster would look like. Of course, you will need an instruction manual to go along with the robot so that your cousin will know how to operate it. Keep the manual short—no more than two pages. It should include step-by-step instructions and may include pictures.

| **Indicators** | **Dimensions** |
|---|---|

### ACTIVITY 1 **Metals that Attract**

| | |
|---|---|
| Predicts based on logical reasoning | S |
| Records data for magnetic items (paper clip, brass fastener, pin) | S |
| Observes that all metal objects are not magnetic | S |
| Infers that all metal objects are not made of the same materials (might say that some are made of iron) | S, C |
| Gives an example of the usefulness of magnets | STS, C |
| Explains how the magnet can be used | C |

### ACTIVITY 2 **Close Encounters**

| | |
|---|---|
| Predicts with logical explanation | S |
| Describes the test in words or drawings | S |
| Performs the test (observation) | S |
| Makes a data table | S |
| Shows reasonable measurements | S |
| Concludes that a magnet does not have to be in contact with an object to attract it | S, C |
| Draws the magnet showing lines of force with limited range | C |

### ACTIVITY 3 **Coming ATTRACTions**

| | |
|---|---|
| Records observations for the three situations | S |
| Concludes that like poles attract and unlike poles repel | S |
| Tells how a magnet is like a force | C |
| Lists two forces | C |
| Tells how the forces affect people or objects | C, STS |

**Suggested Criteria for Writing Prompt:** The instruction manual should be organized with a sequence of logical steps. The instructions should indicate an understanding of magnetic attraction. Pictures are optional (or may be required) but, if used, should fit the logical sequence of steps and aid in understanding how to use the robot.

# Water, Water Everywhere

## CHAPTER 8

## Targeted Process Skills

- observing
- measuring
- inferring
- reasoning
- recording data
- predicting
- drawing conclusions
- researching information

## Rationale

Students can see water as rain, snow, and ice, but evaporation—the changing of water into vapor—is less noticeable. Activities where students observe the disappearance of water from a dish help to reinforce this natural phenomenon. Condensation, the formation of droplets of water on particles in the atmosphere or on surface objects (i.e., dew on grass, frost on a window), allows students to again see water in a visible form. Clouds and fog are visible, but do students realize these are made up of droplets of water? Students should be aware of the water cycle in nature and be able to trace it through its various stages.

Activity 1 allows students to observe the water cycle over time in a closed environment. Because the water cannot leave, it is present in liquid (visible) and vapor (invisible) stages. Students must infer the stages that cannot be seen and draw a logical conclusion about the process as a whole.

In activity 2, students will use the vocabulary of evaporation, condensation, and accumulation to describe the cycle they observed and will tell what precipitation looks like. They will describe the water cycle in words and pictures.

---

### PROJECT 2061 STATEMENTS
#### from *Benchmarks for Science Literacy*

When liquid water disappears, it turns into a gas (vapor) in the air and can reappear as a liquid when cooled, or as a solid if cooled below the freezing point of water. Clouds and fog are made of tiny droplets of water (p. 68).

• • •

The cycling of water in and out of the atmosphere plays an important role in determining climatic patterns. Water evaporates from the surface of the earth, rises and cools, condenses into rain or snow, and falls again to the surface. The water falling on land collects in rivers and lakes, soil and porous layers of rock, and much of it flows back into the ocean (p. 69).

• • •

Offer reasons for their findings and consider reasons suggested by others (p. 286).

• • •

Know why it is important in science to keep honest, clear, and accurate records (p. 287).

---

In activity 3, students will discover that only water in its pure form goes through the water cycle. Salt that is in water is left behind when the water evaporates, condenses, and reaccumulates as liquid on the bottom of the Ziploc bag. Students will relate this phenomenon to what happens in nature and determine that the water cycle is a distilling process. They will also research acid rain as an S-T-S connection. This performance assessment may be linked to units of study dealing with ecosystems, oceans, fresh water, environmental factors, and rain forests.

PERFORMANCE
ACTIVITY

## MATERIALS

❏ graduated cylinders

❏ small plastic cups (3 oz.)

❏ large Ziploc bags

❏ tape

# Water Cycle Observations

## Description of Activity

This activity gives students an opportunity to make observations of a water cycle in action over four days. (You may also want to set up activity 3 at this time. See "Description of Activity" on p. 100). The students will set up closed systems in Ziploc bags and be responsible for observing their own systems throughout the week. The total amount of water measured into the cup should remain the same after all the water has been allowed to evaporate, condense, and collect at the bottom of the Ziploc bag. Data will be recorded and inferences will be made.

## Presenting the Activity

There are a variety of contexts for using this assessment. One example would be as part of a study of the rain forests. Students would learn that these mysterious places occupy about 8 percent of the earth's land surface but house 40–50 percent of the world's plants and animals. Five million species of plants, insects, and animals live there, but only one fourth of these have been identified and studied. About 80 percent of the world's insects live in the tropical rain forests, but many species are disappearing and may never be known to humans. Each year an area of tropical rain forest the size of the state of Indiana is destroyed. The rain forests capture, store, and recycle water, which prevents floods, droughts, and erosion of the soil. In an effort to understand more about the rain forests' contribution to water purification, students may decided to study more about the way that water is cycled through the natural system. (*Rainforests* by Rodney Aldis is an excellent resource for further study on this type of ecosystem.)

# Water Cycle Observations

**What happens to liquid water when it is left in a warm, closed system such as inside a Ziploc bag?**

Tropical rain forests are home to five million species of plants, insects, and animals, many of which will become extinct before humans can even discover them. In this type of ecosystem, water is a plentiful and valuable resource. The rain forests capture, store, and recycle water, which prevents floods, droughts, and erosion of the soil. In an effort to understand more about the rain forests' contribution to water purification, you can study what happens to water in a closed environment.

Using the graduated cylinder, measure 40 ml of water into a small cup. Place the cup in the corner of a Ziploc bag. Tape the cup into place. Zip the bag shut and tape it to a window where it will get sunlight and heat.

Your closed system should look like this:

Predict what will happen to the water inside the cup over time.

I predict _____

because _____

_____

Observe the bag each day for four days and record information on the chart below.

| Date/Time | Observations |
|---|---|
| _____ | _____ |
|  | _____ |
| _____ | _____ |
|  | _____ |
| _____ | _____ |
|  | _____ |
| _____ | _____ |
|  | _____ |

At the end of day four, estimate or measure how much water is at the bottom of the Ziploc bag. Record your observations below.

_____

_____

Based on your observations, infer where the water on the bottom of the Ziploc bag came from.

_____

_____

# Additional Lesson Ideas of Your Own

PERFORMANCE
ACTIVITY 2

**MATERIALS**

# Comparing Water Cycles

## Description of Activity

Students will use the vocabulary related to the water cycle—evaporation, condensation, accumulation, and precipitation—to describe the various phases they observed in the closed system. They will describe the evidence they saw and the inferences they made for each of the stages. Students will use words and pictures to describe the water cycle in the closed system.

## Presenting the Activity

Now that students have observed the water cycle in a closed system and collected data, they can relate this process to what occurs in the natural environment. They will draw and label the cycle as they compare the water cycle in the Ziploc bag with the water cycle in the natural environment.

# Comparing Water Cycles

Relate the process you observed in the closed system to the water cycle in nature. Tell what evidence there is that each process—evaporation, condensation, and accumulation—is occurring.

| Process | Evidence |
|---------|----------|
| evaporation | _____ |
| condensation | _____ |
| accumulation | _____ |

What would the process of precipitation look like in nature's water cycle?

_____

What did it look like in the closed system?

_____

Draw the water cycle below as it occurred in the Ziploc bag. Label the area where each process occurs (evaporation, condensation, and accumulation).

PERFORMANCE
ACTIVITY 3

# A Salty Experience

## Description of Activity

In this activity students will discover that if salt is added to the water, it remains in the cup and does not go through the water cycle. You may want to have students set up this activity at the same time they are doing the first activity or after 1–2 days so that they will not have to wait another week for data. Or you may have students work in pairs where one student sets up a water cycle with plain water and the other student uses salt water. By observing that the salt remains in the cup, students will infer that the water cycle "distills" water, leaving the impurities behind. Students are also asked to make some S-T-S connections about what this means for the earth's supply of fresh water.

## Presenting the Activity

Ask students, "If salt were added to the water in the cup, what would happen to the salt during the water cycle? Would the salt stay with the water through the process or would it remain in the cup?" (If students have not done activity 1, they may not realize that the water will leave the cup.) Then ask the more general question, "What would happen to the water if it were salt water?" Students will predict and give a reason for their predictions. Then, they will collect data over the course of four days. They will observe the water and make inferences about the salt in the cup.

Finally, have students consider how the condition of acid rain occurs. If the water cycle is a distilling process, how does rain get acidic? What effect does acid rain have on buildings, statues, plants, etc.? In the writing prompt, students will research and provide information on acid rain.

## MATERIALS

❏ graduated cylinders

❏ small plastic cups (3 oz.)

❏ Ziploc bags

❏ tape

❏ measuring spoons (metric)

❏ salt

# A Salty Experience

**What would happen if the water in the cup were salt water?**

Repeat the procedure from activity 1, but this time add 5 ml of salt to the 40 ml of water in the cup. Predict what you think will happen to the salt throughout the water cycle.

I predict _____

because _____

Observe the Ziploc bag each day for four days and record your observations on the chart below.

| Date/Time | Observations |
|-----------|--------------|
| _____ | _____ |
|           | _____ |
| _____ | _____ |
|           | _____ |
| _____ | _____ |
|           | _____ |
| _____ | _____ |
|           | _____ |

Estimate how much water is at the bottom of the Ziploc bag.

_____

Is the water in the cup salty or plain? _____

Is the water on the bottom of the Ziploc bag salty or plain? _____

What can you conclude about the water cycle?

_____

_____

_____

_____

How is the supply of fresh water in the world dependent on the water cycle?

_____

_____

_____

_____

Based on what you learned in the activity, describe what the "ideal" quality of rain water would be.

_____

_____

_____

_____

# Water, Water Everywhere

Note that the correct answer for all multiple-choice questions is letter "a." Teachers will need to vary the answers as questions are selected for use.

1. In this diagram representing the water cycle, the process of evaporation is represented by which arrow?
   a. arrow 3
   b. arrow 2
   c. arrow 1

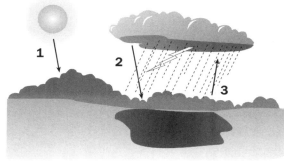

2. In the above diagram, the arrow representing the process of precipitation is
   a. arrow 2
   b. arrow 3
   d. arrow 1

3  In the water cycle, water changes to water vapor through the process of
   a. evaporation
   b. precipitation
   c. adaptation
   d  condensation

4. In the process of evaporation, liquid water is changed to
   a. water vapor
   b. precipitation
   c. clouds
   d. frozen water

5. In the process of condensation, water vapor is changed to
   a. water droplets
   b. a gas
   c. water vapor
   d. evaporated water

6. In the water cycle, water in the form of a gas changes back to its liquid form during the process of
   a. condensation
   b. evaporation
   c. precipitation
   d. accumulation

7. Clouds and fog are made up of droplets of water that form on particles during the process of
   a. condensation
   b. evaporation
   c. precipitation
   d. accumulation

8. Rain, snow, and sleet are forms of
   a. precipitation
   b. condensation
   c. evaporation
   d. clouds

9. The source of energy for the water cycle is the
   a. sun
   b. clouds
   c. atmosphere
   d. water

10. Water in the oceans would not be able to evaporate if it were not for the
    a. sun
    b. salt
    c. air
    d. icebergs

11. Water is placed in two shallow dishes. One dish is placed in a warm spot and one in a cool spot. What is likely to happen to the evaporation rate of the water in these two dishes?
    a. the one in the warm spot will evaporate first
    b. the one in the cool spot will evaporate first
    c. they will evaporate at the same rate
    d. the water will not evaporate from these dishes

12. The water cycle is a continuous process. Which of the following is NOT an example of a cycle?
    a. baking a cake
    b. the seasons
    c. rock formation and break down
    d. stages in the life of a frog

13. All cycles are alike because they
    a. are continuous
    b. have the same parts
    c. grow
    d. end

14. Life cycles and the water cycle are similar because
    a. they have no beginning and no end
    b. they both deal with the universe
    c. one is living and one is non-living

# Acid Rain

Through the activities, you have learned that the water cycle distills water, that is, it separates water naturally from impurities. If this is so, how does rain water become acid rain?

Your task is to research acid rain. Your report should consist of several paragraphs that answer the following questions:

- how does rain becomes acid rain?

- what effect does acid rain have on buildings, statues, plants, etc.?

- what can be done to minimize or eliminate acid rain?

- should people be concerned about acid rain and why?

You should use at least two different sources of information in writing your report. You may use written resources, video programs, computer-accessed information, and/or human resources. Give a complete reference for each source used.

| **Indicators** | **Dimensions** |
|---|---|

### ACTIVITY 1   **Water Cycle Observations**

| | |
|---|---|
| Measures water into the cup | S |
| Makes a prediction | S |
| Gives an explanation | S |
| Records observations for four days | S |
| Estimates/records amount of water in the Ziploc bag | S |
| Infers how the water got on the bottom of the bag | S |

### ACTIVITY 2   **Comparing Water Cycles**

| | |
|---|---|
| Describes evidence for each of three processes | S, C |
| Tells what precipitation looks like in nature | C |
| Tells what precipitation looks like in a closed system | C |
| Draws the water cycle | C |
| Labels each of three processes | C |

### ACTIVITY 3   **A Salty Experience**

| | |
|---|---|
| Measures water and salt into the cup | S |
| Makes a prediction | S |
| Gives an explanation | S |
| Records information for four days | S |
| Estimates amount of water in the Ziploc bag | S |
| Identifies and records salty water | S |
| Identifies and records plain water | S |
| Writes a conclusion about the process | S, C |
| Describes importance of water cycle | C, STS |
| Describes the "ideal" quality of rain water | C, STS |

**Suggested Criteria for Writing Prompt:** The paragraphs should adequately answer the four questions and include the required sources of information. Dimensions of language arts may be assessed, such as the mechanics of writing a paragraph.

# A Wholesome Partnership

## CHAPTER 9

## Targeted Process Skills

- observing
- communicating
- measuring
- reasoning
- inferring
- estimating
- collecting data
- drawing conclusions
- problem solving
- comparing data
- developing formulae

### PROJECT 2061 STATEMENTS
#### from *Benchmarks for Science Literacy*

No matter how parts of an object are assembled, the weight of the whole object is always the same as the sum of the parts; and when a thing is broken into parts, the parts have the same total weight as the original thing (p. 77).

• • •

Mathematical statements can be used to describe how one quantity changes when another changes (p. 219).

## Rationale

This performance assumes that students have had experience using a balance and set of weights to find the mass of a variety of objects and have had experience with whole/part relationships. They should also have experience working with and constructing data tables.

Students will be asked to find the mass of the whole object (banana) as well as each of the two component parts. Students should be able to explain any difference when the weight of the two parts (edible and inedible) does not equal the weight of the whole. It is important that students defend the fact that the mass of the whole object is always the same as the sum of the parts. Any discrepancy should be attributed to such variables as the accuracy of the balance, types of weights used, precision of the procedure and researcher, data recording techniques, etc.

If discrepant results are found (the whole does not equal the sum of the parts), students should show an eagerness to redo the activity for more accuracy rather than be satisfied with results that do not show that they understand the concept being assessed.

In activity 1, students will make estimates, record group data, find averages, and draw conclusions. Finally, they will be asked to develop a formula based on the results of the measurement and defend the concept that the whole is equal to the sum of its parts.

In activity 2, students will make an estimate about the percentage of the edible portion in very ripe, green, small, or large bananas or in some other variety of banana such as plantains or miniature bananas. Students will have an opportunity to make their own data tables, record data, draw conclusions, and compare findings with the results of activity 1.

An optional activity could ask students to research in groups such information as the history of bananas, the geographical sources of bananas, the nutritional value of bananas, and/or the banana plant. Each research group should design an instructional strategy for teaching what they learned to the other groups. Creative presentations might include performing a rap, playing a game, performing a play, or displaying posters, charts, graphs, or other visuals to accompany a verbal presentation.

Teachers may wish to develop criteria and to design an observation checklist for skills that will be assessed during this activity, such as working with a group, researching information, staying on task, recording information, taking part in the performance, showing a positive attitude, and taking initiative.

The activities in this performance assessment are based on an activity called The Big Banana Peel developed by the AIMS Education Foundation, P.O. Box 8120, Fresno, CA 93747.

PERFORMANCE ACTIVITY

# Sum-thing Fruitful

## Description of Activity

Students are each given a banana and asked to examine the whole structure and to estimate what percentage of the total mass of the banana the edible portion represents. Students may discuss ideas but should record individual estimates. After finding the mass of the whole object and its edible and inedible parts, students will record data, make graphs, and explain their findings.

## Presenting the Activity

Tell students they will be working with a banana to determine what percent of the whole the edible portion represents. Operationally define "edible portion" as the white part that separates naturally from the outer skin. Instruct them to read the directions and complete the data sheet, making sure to write responses in the appropriate spaces.

## MATERIALS

❑ balances
❑ sets of weights
❑ paper towels
❑ bananas

# Sum-thing Fruitful

**What percentage of a banana is edible?**

Name: _____

Date: _____

I think _____

because _____

Find the mass of the whole banana, the edible portion, and the peeling and record the information on the data table below.

| Object | Mass (in grams) |
|--------|-----------------|
| Whole Banana | _____ |
| Edible Portion | _____ |
| Peeling | _____ |

| Ratio (edible/whole) | Decimal | Percent |
|----------------------|---------|---------|
| _____ | _____ | _____ |
| **Ratio (peeling/whole)** | **Decimal** | **Percent** |
| _____ | _____ | _____ |

What did you observe?

_____

_____

_____

_____

What can you say about the relationship between the parts of the banana and the whole banana?

_____

_____

_____

_____

What formula (using decimal or percent) would represent the edible portion of the banana?

**Formula for the Edible Portion =** _____

How do your results compare with the results of others? Collect data from five groups or individuals and record the data below:

| Banana # | Mass of Whole Banana (g) | Mass Edible (g) | Mass of Peeling (g) | Percent Edible (%) |
|----------|-------------------------|-----------------|---------------------|---------------------|
| 1 | _____ | _____ | _____ | _____ |
| 2 | _____ | _____ | _____ | _____ |
| 3 | _____ | _____ | _____ | _____ |
| 4 | _____ | _____ | _____ | _____ |
| 5 | _____ | _____ | _____ | _____ |

Find the average percent edible for the six bananas.

**Average Percent =** _____ %

# Additional Lesson Ideas of Your Own

## PERFORMANCE ACTIVITY 2

### MATERIALS

❏ balances

❏ sets of weights

❏ paper towels

❏ bananas different from those used in activity 1 (common bananas that are very ripe, green, miniature, or extra large or a different type banana)

# Going Bananas

## Description of Activity

Students are asked to reflect on what they found in activity 1 and to consider what percent of the whole the edible portion will be in another banana that has different characteristics or is a different type. Students will estimate the percent for the new banana, find the mass of the whole and the parts, construct a data table, and draw conclusions. Students will be asked to develop a formula for the new relationship and to compare findings from activities 1 and 2.

## Presenting the Activity

Ask students if they think the percent of the edible portion of a banana is the same when the conditions are different, such as when the banana is very ripe or when it is very green. Give students another type of banana or one that has different characteristics than the one used in activity 1. Tell them to follow the directions on the student sheet, making sure to record the data for each question in the appropriate space.

# Going Bananas

**Will the percentage of the edible part be the same for a banana with different characteristics than the one you investigated in activity 1?**

I think _____

because _____

Based on what you learned in activity 1, you will make a statement about the relationship between the whole fruit and the edible portion for a banana that is different from the one used in activity 1.

Record the new banana's characteristics:

_____

List one way it is like the banana in activity 1:

_____

List one way it is different from the banana in activity 1:

_____

Make an estimate of the percent of the edible portion of this banana (or type of banana):

**Percent Edible =** _____

Now find the mass of the whole object, the edible portion, and the peeling. Calculate the percent of the edible portion. Make a data table below and record the data.

What conclusion can you draw from your data?

_____

Show a formula (percent edible) for this new banana.

_____

How do your findings compare with what you found in activity 1?

_____

_____

_____

_____

# A Wholesome Partnership

Note that the correct answer for all multiple-choice questions is letter "a." Teachers will need to vary the answers as questions are selected for use.

1. Sue and her dad went to the store and bought a turkey for Thanksgiving dinner. The turkey weighed 5 kilograms. If the bones and inedible parts of the turkey weighed 2 kilograms, the edible portion that was left weighed
   a. 3 kilograms
   b. 5 kilograms
   c. 2 kilograms
   d. 8 kilograms

2. Mr. Jones wanted to buy large boulders as decorations for his yard. At the landscaping store he found three boulders, with masses of 30 lbs., 36 lbs., and 44 lbs. The total weight of the three boulders is
   a. 110 lbs.
   b. 100 lbs.
   c. 120 lbs.
   d. less than 100 lbs.

3. Paul found that the mass of a large apple was 500 grams. He divided the apple into four parts and found that three had the following masses: 100 grams, 150 grams, and 125 grams. The mass of the fourth piece should be
   a. 125 grams
   b. 150 grams
   c. more than 150 grams
   d. less than 125 grams

4. Suzy was taking a box with a glass bowl to the post office to be weighed. On the way to the office, she dropped the box and heard a crash. The broken bowl in the box weighs
   a. the same as the bowl when it was whole
   b. less than the bowl when it was whole
   c. more than the bowl when it was whole
   d. the same as the box

5. The candy store was having a sale on licorice at two dollars per pound. Diann wanted to buy some licorice but only had one dollar. She found Tony, who also had one dollar. Together Diann and Tony could buy how much licorice?
   a. a pound
   b. less than a pound
   c. more than a pound
   d. two pounds

6. You find the weight of a large rock. Then you break it apart into five smaller pieces. You weigh each piece and add the weights of pieces. The total weight of the pieces will be
   a. the same as the weight of the large rock
   b. less than the weight of the large rock
   c. more than the weight of the large rock
   d. similar to the weight of one piece

7. Franz has a bag of five apples. He finds the weight of the bag with the apples. Then he weighs each of the apples and the bag and adds their weights. What should Franz discover about the apples?
   a. The weight of five apples plus the bag is the same as the total weight of the bag
   b. The weight of five apples is the same as the total weight of the bag
   c. The individual apples weigh more than the total bag
   d. The individual apples plus the bag weigh less than the total bag

8. A plant weighs 1 kilogram. John takes the plant apart and weighs each individual part. Then he adds the weights of the parts and finds that they weigh
   a. 1 kilogram
   b. less than 1 kilogram
   c. more than 1 kilogram
   d. totally different than the whole plant

# A Helpful Partnership

Jacques and Sylvia are doing activity 1. They discover that the mass of the edible portion of their banana plus the mass of the peeling does NOT equal the mass of the whole banana. You have been asked by the teacher to explain to them why results like this might occur. What would you tell them? Give your answer in a paragraph or two. In your explanation, provide:

- two or more possible reasons for what they found

- an explanation shown mathematically

- a plan for getting more accurate data in making comparisons

| **Indicators** | **Dimensions** |
|---|---|

### ACTIVITY 1  **Sum-thing Fruitful**

| | |
|---|---|
| Predicts based on logical reasoning | S |
| Measures the mass of the whole banana | S |
| Measures the mass of the edible portion | S |
| Measures the mass of the peeling | S |
| Calculates the percent edible (may add points for each mathematical calculation) | S |
| Makes an observation of data | S |
| States a relationship | S, C |
| Develops a formula based on data | S, C |
| Completes data table with information | S |
| Finds the average percent | S |

### ACTIVITY 2  **Going Bananas**

| | |
|---|---|
| Makes a prediction | S |
| Records information | S |
| Lists one similarity | C |
| Lists one difference | C |
| Makes an estimate of the percent edible | S |
| Measures the mass of the whole object | S |
| Measures the mass of the edible portion | S |
| Measures the mass of the peeling | S |
| Draws a conclusion | S |
| Develops a formula based on data | S |
| Compares findings | C |

**Suggested Criteria for Writing Prompt:** The paragraphs should contain a minimum of two inferences, an explanation shown mathematically, and a reasonable plan for data collection, which could be as simple as conducting additional, similar investigations. Dimensions of language arts may also be assessed, such as the mechanics of writing a paragraph.

# A-W-L for One and One for A-W-L
## (Air-Water-Land)

## CHAPTER 10

## Targeted Process Skills

- observing
- problem solving
- inferring
- communicating
- measuring
- collecting data
- predicting
- drawing conclusions
- comparing
- reasoning

## Rationale

In this performance, students will take on the role of environmental investigators for the Community Environmental Task Force. Students will be allowed to choose to investigate air, water, or land in the local environment for visible signs of pollution. This performance should follow a study of pollution where students realize that pure air, clean water, and unpolluted land are precious and valuable resources, necessary for the continued survival of living things.

Three different investigative activities will be presented. The first studies air; the second samples water; the third surveys land. Directions for sampling and investigating are given, however, such factors as the places for sampling air, water, and land will be determined by students. The quality of the investigation, in some cases, will be determined by the choices that students make. Thus, as part of the summary report that students present in activity 4, they will self-assess the choices they made.

Allow students to select activity 1, activity 2, or activity 3 as their investigative activity. All students should do activity 4, which asks them to prepare a short presentation using a visual

---

### PROJECT 2061 STATEMENTS
#### from *Benchmarks for Science Literacy*

Scientists do not pay much attention to claims about how something they know about works unless the claims are backed up with evidence that can be confirmed . . . with a logical argument (p. 11).

• • •

Statistical predictions are typically better for how many of a group will experience something than for which members of the group will experience it—and better for how often something will happen than for exactly when (p. 227).

• • •

Students should keep records of their investigations and observations and not change the records later (p. 286).

• • •

Students should offer reasons for their findings and consider reasons suggested by others (p. 286).

---

aid to describe their investigation and findings. By investigating signs of pollution in the immediate environment, students will realize that they can be personally affected by pollution.

**MATERIALS**

❏ one transparency or sheet of plastic per student

❏ petroleum jelly

❏ three plastic baggies per student

❏ string

❏ masking tape

❏ hand lenses

# AIRing Our Differences

## Description of Activity

Students will investigate air samples to determine the types and amount of particles that they find.

## Presenting the Activity

Tell students they are environmental investigators for the Community Environmental Task Force and have been asked to study the quality of the air in their community. They should think about what factors affect air purity. There are many ways that particles can get into the air; students should think about some of the particles they have seen mixing into the air. These might include airplane, bus, automobile, and motorcycle exhaust; smoke from factory, office, or home chimneys; smoke from burning leaves; excessive amounts of dust from fields; and pollen from plants.

Students will cut a sheet of plastic into four sections. They will punch a hole at the top of three of these pieces and thread a 10–12 cm piece of string through and tie the string so that it can be used to hang the particle detector from a nail or branch. Next, they will smear petroleum jelly on one side of each piece. Students will select three sites to test. These should be places where students think there might be a lot of particulate matter in the air, but they should also be places that are easily accessible. They might select a section in their garage to trap particles from automobile exhaust, a dusty place, or an area near the school that buses frequent. The particle detectors should be placed where they will not be disturbed for at least two to three days.

After a few days, students will collect the particle detectors and put each one into a separate plastic sandwich bag that is labeled with the location of the site it came from. Students will examine the particle detectors with a hand lens to determine the types and amounts of particulate matter that are present. Before doing activity 4, they should answer the questions on the student sheet concerning sources of particles, possible ways to reduce particles in the air, and the importance of clean air.

Name: _____

Date: _____

# AIRing Our Differences

**What places in your community have the most air pollution?**

You are asked to be a member of the Air Quality Team of the Community Environmental Task Force. Your job is to sample three sites in your community for the types and amount of particles found in the air. You will prepare a report and give a presentation to community leaders, who are concerned about the health hazards present in the local environment.

One type of pollution found in the air is particulate matter. Particles may be present from dust, smoke, or chemicals. Particle detectors can be made by taking a sheet of plastic and cutting it into four equal parts. Punch a hole at the top center of three pieces and tie a ten-centimeter piece of string through it. Coat the three pieces on one side with petroleum jelly. Now the particle detectors are ready to hang or tape in places where particles can be collected from the air.

Think about places in your community that might have a lot of particles in the air. Your task is to sample some of the worst examples of pollution that exist in your local environment. Decide on three places that would give you a good sample of the air in your community.

Identify the three places you decide to sample and tell why you chose those places.

**Place**                                      **Reason for Decision**

1._____          _____

2._____          _____

3._____          _____

Predict which of the three places you expect to find the most particles. Tell why you think it will have the most.

I predict _____

because _____

Hang or tape the particle detectors at the three sites and wait two or three days before removing them. Remove the particle detectors by placing each in a small baggie that is labeled with the name of the location. For each particle detector, record the requested information on the table below (use a magnifier to view the particles).

| Place | Amount of Particulate Matter (see scale below) | Texture/ Shapes/Sizes |
|---|---|---|
| 1._____ | _____ | _____ |
| 2._____ | _____ | _____ |
| 3._____ | _____ | _____ |

**Scale**

LOW pollution:        a few particles
MEDIUM pollution:    concentrated spots or light film over all
HEAVY pollution:     more spots or dark film over all

What do you think caused the particles to be in the air?

_____

_____

_____

Can you suggest one way that the amount of particulate matter in the air can be reduced?

_____

_____

_____

Why is it important to have clean air?

_____

_____

_____

# Wat-er We Finding in Our Water?

## Description of Activity

Students will take three samples of water from the immediate environment to investigate the presence of particles.

## Presenting the Activity

Tell students they are environmental investigators for the Community Environmental Task Force and have been asked to study the quality of the water in their community. There are various ways that particles, chemicals, and unwanted materials can get into the water; students should think about some of the factors that affect water purity.

Students will identify three areas in the community to study. They may select samples from bodies of fresh water such as ponds or streams. They may also consider puddles, collected rain water, well water, tap water, river water, water from a golf course, or other sources. Students should use jars to collect approximately the same amount of water from each site. A simple filter system can be constructed by fitting a coffee filter inside the soup cans in which holes have been punched. Students will observe the color and smell of the sample and then pour the water through the filter, making sure that the water drains into a container or a sink. The filter paper will trap any sizable particles that are in the water. Using a hand lens, students can observe the particles in the filter and record data on the data table. Have students answer the reflection questions at the end of the activity before doing activity 4.

## MATERIALS

❑ coffee filters
❑ small metal soup cans in which about a dozen nail-size holes have been punched in the bottom
❑ hand lenses
❑ collecting jars (baby food or other small jars)

# Wat-er We Finding in Our Water?

**What places in your community have the most water pollution?**

You are asked to be a member of the Water Quality Team of the Community Environmental Task Force. Your job is to sample three sites in your community for the types and amount of particles found in the water and the general color and smell of the water. You will prepare a report and give a presentation to community leaders, who are concerned with the health hazards present in the local environment.

One type of pollution that is found in water is particulate matter. Particles may be present from dust, smoke, or chemicals. Samples of water can be taken from various places in the community and then filtered to separate the larger particles that are present.

Think about places in your community that might have a lot of particles in the water. Your task is to sample some of the worst examples of pollution that exist in your local environment. Decide on three places that would give you a good sample of the water in your community.

Identify the three places you decide to sample and tell why you chose those places.

**Place**                                      **Reason for Decision**

1. _____      _____

2. _____      _____

3. _____      _____

Predict which of the three places you expect to find the most particles in water. Tell why you think it will have the most.

I predict _____

because _____

Visit the three sites. At each one, fill a collecting jar with water. Get the same amount of water from each of the three sites. Tape a label on each jar that identifies the location and the area within the location from which the water was taken (e.g., the local pond—water from the shoreline about 4 cm under the surface).

After you have three samples, construct a filtering system to test the water. Take a metal can that has holes punched in the bottom. Carefully place a coffee filter into the can to cover the holes, keeping the opening as large as possible. Before you pour the water into the filter, observe its color and smell. Record the data on the chart on the next page.

Over a sink or beaker, pour the contents of one collecting jar into the can so that the water goes through the coffee filter and out the bottom of the can. Remove the filter and label it with the location of the site. Do this for the other two samples so that you have three filters with labels to study. Now, open each of the filters and, using a hand lens, observe any particles that are present.

Record your general observations.

_____
_____
_____
_____
_____
_____
_____
_____
_____
_____
_____

For each water sample, record the requested information on the table below.

| Sample/Location | Color/Smell | Amount (use scale below) | Description of Particles |
|---|---|---|---|
| 1._____ | _____ | _____ | _____ |
| 2._____ | _____ | _____ | _____ |
| 3._____ | _____ | _____ | _____ |

**Scale**

LOW pollution:      a few particles

MEDIUM pollution:   concentrated spots or light film over all

HEAVY pollution:    more spots or dark film over all

What do you think caused the particles to be in the water?

_____

_____

_____

Can you suggest one way that the amount of particulate matter in water can be reduced?

_____

_____

_____

Why is it important to have clean water?

_____

_____

_____

# Every Litter Bit Hurts

## Description of Activity

Students will investigate land areas in their community to determine the types and amounts of litter that they find.

## Presenting the Activity

Tell students they are environmental investigators for the Community Environmental Task Force and have been asked to study the quality of the land in their environment. They should think about what factors affect the amount of unwanted litter found on the land. Students will identify three sites to investigate. These should be safe places to visit and/or be places that they can go accompanied by a parent or other adult (i.e., a local park, school grounds, home yard, areas near home or school, areas along streets, and malls). Students should think about the ways that litter can get to the various places they are investigating. They will examine the areas and look for evidence of litter, describe the types and amounts of litter found in each of the areas, and determine whether the amount of pollution/litter was low, medium, or high using the scale on the student sheet. They should answer the questions concerning sources of litter, possible ways to reduce litter, and the importance of litter-free land before doing activity 4.

**MATERIALS**

❑ notebooks
❑ pencils

# Every Litter Bit Hurts

**What places in your community have the most litter?**

You are asked to be a member of the Land Quality Team of the Community Environmental Task Force. Your job is to sample three sites in your community for the types and amount of litter found on the ground. You will prepare a report and give a presentation to community leaders, who are concerned with the health hazards present in the local environment.

One type of pollution that is found on the ground is litter. Litter is usually defined as paper, plastic, glass, wood, or other items that are not a natural part of the environment. Litter may be brought into an area by animals, wind, or humans. Think about places in your environment that might have a lot of litter on the land. Your task is to sample some land areas in your local environment and survey them for the types and amount of litter that are present.

Identify the three places you decide to sample and tell why you chose those places.

**Place**                         **Reason for Decision**

1._____    _____

2._____    _____

3._____    _____

Predict which of the three places you expect to find the most litter. Explain why you think it will have the most.

I predict _____

because _____

For each area, record the requested information on the table below.

| Location | Amount of Litter (use scale below) | Types of Litter |
|----------|-------------------------------------|-----------------|
| 1 _____ | _____ | _____ |
| 2. _____ | _____ | _____ |
| 3. _____ | _____ | _____ |

**Scale**

LOW pollution:      a few pieces of litter

MEDIUM pollution:   litter obvious; several different types

HEAVY pollution:    lots of litter; various types; area looks cluttered/messy/overrun

What do you think caused the litter to be in the area?

_____

_____

Can you suggest one way that the amount of litter can be reduced?

_____

_____

Why is it important to have a clean, litter-free environment?

_____

_____

# Additional Lesson Ideas of Your Own

## PERFORMANCE ACTIVITY 4

# Preparing and Giving an Environmental Task Force Report

### Description of Activity

As members of the Community Environmental Task Force, students will prepare to give a presentation on their findings to the community. You will decide how much time each person will have for their presentation. Each student will make a visual aid to display their findings and present data, make inferences about the quality of the environment based on the findings, and offer suggestions for reducing the amount of pollution in the environment.

### Presenting the Activity

Tell students that after investigating the community for signs of pollution, they have a responsibility to share their findings, inferences, and suggestions for improving the quality of the environment in the community. Their presentations should include the criteria listed on the student sheet.

### MATERIALS

❑ poster board, crayons, markers, etc., for making visual aids

Name: _____

Date: _____

# Preparing and Giving an Environmental Task Force Report

**How do members of Environmental Task Forces share the results of their investigations with other members of the community?**

You will give a presentation to the community (represented by the rest of the class) on the findings of your investigation of pollution in air, in water, or on land. You will have _____ minutes to share your findings.

Prepare a presentation that includes the following:

1. A complete description of your project—identifying the type of pollution you studied and the three sites you chose to study.

2. A visual aid (poster, mobile, pictures, etc.) to display information, data, drawings of your sites, results, or anything else that would be helpful in presenting your project.

3. Findings of your investigation.

4. Inferences about what might have caused the pollution.

5. Evaluation of
   - your choices of sites for investigation (Were they good places? Did you find what you were looking for in these places? Were there other places you wish you had studied instead?)
   - the way you carried out the investigation (Were you careful with the materials you used? Did you follow directions? Did you study the samples carefully enough to get data? Were there other things you could have done to get better results?)
   - the data you collected (Did you get what you thought you would get? Why or why not? If you did this again, do you think you would get different results? If you did this again, what would you do differently?)

6. Suggestions to the community for improving the quality of their environment and suggestions for further investigations.

| Indicators | Dimensions |
| --- | --- |

## ACTIVITIES 1, 2, AND 3

| | |
| --- | --- |
| Identifies three places and gives reasons for decisions | C, S |
| Predicts place of greatest pollution | S |
| Explicates reasons | C |
| Records information on data table | S |
| Makes an inference | S |
| Suggests ways to reduce pollution | C, STS |
| Describes importance of clean environment | C, STS |

## ACTIVITY 4  **Environmental Task Force Report**

| | |
| --- | --- |
| Gives a description of project | C |
| Makes a visual aid appropriate to presentation | S |
| Uses the visual aid to enhance presentation | C |
| Reports findings of investigation | S |
| Makes inferences about what might have caused pollution | S |
| Evaluates choice of sites | S, C |
| Evaluates how investigation was carried out | S, C |
| Evaluates data collected | S, C |
| Makes suggestions for improving quality of environment | C, STS |
| Makes suggestions for further investigations | C |

# The Mysterious Package

## CHAPTER 11

## Targeted Process Skills

- observing
- measuring
- collecting data
- classifying
- inferring
- communicating

## Rationale

This performance consists of one substantial activity that gives students an opportunity to investigate firsthand the diet of a barn owl. Students should have had some prior instruction dealing with energy levels for living things and food chains.

Barn owls are expert hunters with specially designed eyes, ears, claws, and wings to help in catching prey. Owls often swallow their prey whole. They cannot digest bones, fur, and feathers, so these are clumped together in the stomach and regurgitated. When owl pellets are carefully pulled apart, they yield bones and skulls of rodents and/or birds. When the activity is presented in a positive manner, it is an excellent, high-interest study of a food chain. Students are intrigued by what they find!

Ultimately, each student should find at least one animal that he or she can identify from pictures on the charts showing the prey. Students can reconstruct skeletons of the prey. They can draw food chains by researching the types of food that are eaten by the prey. Most of the prey are herbivores, allowing students to trace the food chain back to plants and further to the ultimate source of energy, the sun.

## PROJECT 2061 STATEMENTS

from *Benchmarks for Science Literacy*

A great variety of living things can be sorted into groups in many ways using various features to decide which things belong to which group (p. 103).

• • •

Almost all kinds of animals' food can be traced back to plants (p. 119).

• • •

Some source of "energy" is needed for all organisms to stay alive and grow (p. 119).

• • •

Two types of organisms may interact with one another in several ways: They may be in a producer/consumer, predator/prey, or parasite/host relationship (p. 117).

• • •

Tables and graphs can show how values of one quantity are related to values of another (p. 218).

• • •

The graphic display of numbers may help to show patterns such as trends, varying rates of change, gaps, or clusters (p. 224).

PERFORMANCE
ACTIVITY

# Large Things Come in Small Packages

## Description of Activity

Students will investigate an owl pellet. They will measure and examine it before dissecting it to examine its valuable contents. They will carefully observe and classify the bones and skulls so that they can match their finds with types of bones and skulls found on the identification charts. The best part of this activity is that each pellet is different, adding an element of curiosity. Students may work individually or in pairs. They will make individual bar graphs but may also make a class graph of the prey found. Be sure to allow plenty of time for finding, sorting, and reconstructing. This activity may need to be done in several sessions.

## Presenting the Activity

Scientists (zoologists or field biologists) who are interested in learning about animals, their habitats, food, behaviors, and such are always intrigued by evidence they can experience firsthand. Explain to students that owl pellets will provide them with the opportunity to investigate the diet of the barn owl. Owl pellets are extraordinary since they provide so much information about the barn owl's diet and the population of animals in the areas where the pellets were found. (Commercially purchased pellets have been sprayed and wrapped, but otherwise they are very much like they were when collected from the barn where the owls roosted.) Students can observe the specimens and infer the total number and types of prey. If students are given a positive introduction to the investigation, they will be eager to investigate the pellets.

Students should measure and describe the pellets before dissecting. They should have access to sorting sheets and charts to assist in identifying bones and skulls. Egg cartons may be used to sort bones. A culminating activity would be to reconstruct the skeleton of one or more of the prey. Students may need to work in pairs or triads in order to have all the bones they need for one good specimen.

There are a number of social issues related to the study of owls. There is a controversy in the northwest dealing with the habitat of the spotted owl; the logging industry has seriously threatened the owl's habitat. In other areas, harsh weather, shortage of food, and habitat loss are problems for the owls. Because these issues are always changing, it is suggested that students watch for them in current events or research them for an S-T-S connection.

### MATERIALS

- ❏ owl pellets and prey charts (pellets and charts can be ordered from Pellets, Inc., 3004 Pinewood, Bellingham, WA 98225)
- ❏ tweezers
- ❏ probes (or toothpicks)
- ❏ paper towels
- ❏ egg cartons (optional)

# Large Things Come in Small Packages

**Can a rodent or bird fit inside an owl pellet?**

Measure and record the length of your owl pellet.

_____

Measure and record the distance around the center of your owl pellet.

_____

Describe the pellet (without the wrapper).

_____

_____

_____

Carefully separate the bones from the fur, feathers, and other items in the pellet. Clean the bones and/or skulls and sort them by type using a sorting sheet, egg carton, or sections on a piece of paper.

Sketch each type of bone you found and record how many of each type were in the pellet.

**Type** (sketch)                                              **Number**

Make a bar graph of the types and number of bones in your owl pellet.

How many skulls did you find? _____

If a bird consumes two prey each day, how many will it eat a year? _____

Consider your findings as well as those of others in the class. What can you infer about the population of prey in the area where the pellets came from?

_____

_____

_____

_____

Below, draw a food chain with the barn owl as the top predator. Be sure to trace back to the source of energy for the entire food chain. Label all components of the food chain.

# The Mysterious Package

Note that the correct answer for all multiple-choice questions is letter "a." Teachers will need to vary the answers as questions are selected for use.

1.  Consider the food chain:
    PLANTS ⟶ GRASSHOPPER ⟶ TOAD ⟶ RACCOON

    Which is the producer?
    a.  plants
    b.  grasshopper
    c.  toad
    d.  raccoon

2.  The organisms that are able to make their own food are called
    a.  producers
    b.  consumers
    c.  decomposers
    d.  herbivores

3.  Green plants are very important on the food chain because they are
    a.  producers
    b.  consumers
    c.  decomposers
    d.  herbivores

4.  Animals that eat plants are called
    a.  herbivores
    b.  carnivores
    c.  top predators
    d.  producers

5.  Consider the food chain:
    CARROT ⟶ RABBIT ⟶ HUMAN

    The herbivore is the
    a.  rabbit
    b.  carrot
    c.  human
    d.  top predator

6. Which of the following would be the lowest consumer on the food chain?
   a. herbivore
   b. carnivore
   c. top predator
   d. plant

7. The energy for a food chain is provided by
   a. the sun
   b. plants
   c. herbivores
   d. carnivores

8. Plants get the energy to produce food from
   a. the sun
   b. photosynthesis
   c. animals
   d. soil

9. A food chain would not exist without
   a. sunlight
   b. predators
   c. carnivores
   d. sand

# The Changing Environment

A community of owls lives in an old red barn near a small but growing town in the state of Washington. The owls have been there for over fifteen years, but it seems that food is harder to find every year. Numerous subdivisions and strip malls have been built over the years, bringing people closer and closer to the barn. You have become aware that next year the barn may be taken down and the land sold for houses.

The Owl Preservation Society has asked your help in designing a half-page ad they can put into the local paper. The ad should include one or more drawings as well as text. The Preservation Society hopes that the citizens of the area will realize the importance of owls to the environment and some possible effects of the loss of owls to their area.

Use the information you found in your "mysterious package" and what you know about owls to design the half-page ad. Make your ad campaign as convincing as possible. The owls are depending on you to help them keep their homes.

**Indicators**                                                              **Dimensions**

## Large Things Come in Small Packages

| | |
|---|---|
| Measures the length of the pellet | S |
| Measures the circumference of the pellet | S |
| Describes the pellet | C |
| Sorts bones by type | S |
| Draws each type | S |
| Counts each type and provides numbers | S |
| Makes a bar graph | S |
| Calculates number or prey eaten in a year | S |
| Infers the total number and types of prey | S |
| Draws a food chain including owl and prey | C |
| Shows what food prey would eat | C |
| Shows source of energy—the sun | C |

**Suggested Criteria for Writing Prompt:** The advertisement should be one-half page and should include a minimum of one drawing and text to describe the message. The ad should address the importance of owls to the environment and identify a minimum of one effect (inferred) that relates to the loss of owls to the area. Information uncovered in the mystery package should be incorporated into the ad.

# Scientific Investigations

## CHAPTER 12

## Targeted Process Skills

- observing
- operational questioning
- collecting data
- inferring
- hypothesizing
- analyzing data
- drawing conclusions
- classifying

## Rationale

Not all of what we learn comes through controlled experiments. Many great science units are made up of a series of investigations where key questions are asked, manipulative activities are presented or suggested, data are collected, and experiences and/or results are processed through discussion. Through the investigative process, more questions emerge that prompt additional investigations, and the process of discovery learning is enhanced. Investigative activities are critical to the primary- and middle-grade science program since they provide an experience base for constructing meaning, promote the development of skills and higher-order thinking, and provide a rich environment for the development of science literacy.

Throughout the upper elementary grades and high school, the questions students ask or are asked should become more sophisticated. Moreover, students should know how to set up an experimental situation and conduct a controlled experiment where they investigate the effect of one variable on another.

## PROJECT 2061 STATEMENTS

### from *Benchmarks for Science Literacy*

Scientists do not pay much attention to claims about how something they know about works unless the claims are backed up with evidence that can be confirmed . . . with a logical argument (p. 11).

• • •

Scientists differ greatly in what phenomena they study and how they go about their work. Although there is no fixed set of steps that all scientists follow, scientific investigations usually involve the collection of relevant evidence, the use of logical reasoning, and the application of imagination in devising hypotheses and explanations to make sense of the collected evidence (p. 12).

• • •

If more than one variable changes at the same time in an experiment, the outcome of the experiment may not be clearly attributable to any one of the variables. It may not always be possible to prevent outside variables from influencing the outcome of an investigation, but collaboration among investigators can often lead to research designs that are able to deal with such situations (p. 12).

• • •

Thinking about things as systems means looking for how every part relates to others (p. 265).

Prior to this assessment, students should have had experience identifying types of variables and carrying out experiments where they have the opportunity to test the effects of one variable on another. In activity 1, students will make a simple kite and determine the effect of the variables—point of attachment of the string to the kite and the length of the kite's tail—on the performance of the kite (the way it flies). Activity 2 allows students to make decisions about the type of tail they will test in order to decide which design allows the kite to fly best. Students will have the opportunity to be creative and imaginative in their decision making. The first kite activity introduces the more complex second kite activity.

The rubric for this assessment deals with the indicators of dimensions in conducting a controlled experiment. If the context of the assessment is building kites, applications of knowledge about kite aerodynamics can be built into the activities and these indicators can be added to the scoring rubric.

## PERFORMANCE ACTIVITY

**MATERIALS**

❑ 8.5 in. x 11 in. sheet of paper per person

❑ one plastic straw per person

❑ transparent tape

❑ metric rulers

❑ kite string

❑ paper punch

❑ kite directions

# Go Fly a Kite

## Description of Activity

Students will construct simple straw kites (see directions on pages 150–152). All students will build kites to the same specifications in this activity, including the type and length of kite tails. Students will determine which point of attachment—front point (6 cm from front), midpoint (7 cm from front), or rear point (8 cm from front)—gives the best flight performance. Because all kites are identical, results of the experiments should be similar. However, uncontrolled variables such as slight differences in the process of making the kites and conditions for flying the kites (wind conditions, position of students in relation to the wind, buildings and obstructions, etc.) may account for differences in results.

Students should keep a detailed account of what they test, where they test, and the results of trials. If students are allowed to give verbal explanations of their experiences, teachers can learn a great deal about a student's understanding of the experimental process. Teachers may determine that specific habits fit the performance assessment, such as using logical reasoning skills, valuing investigations, recognizing the importance of record keeping, etc. Teachers may develop a checklist describing such habits of mind and observe student behaviors during the investigations. Behaviors relating to selected habits of mind may not be observed for all students during a performance task; teachers may wish to allow themselves more time to observe and record data for each student.

## Presenting the Activity

Tell students that the local park district is going to offer a summer program for young children that will teach them about kite aerodynamics and provide an opportunity to make and fly simple kites. They have asked students to investigate the conditions under which the kite will fly best and to determine the position of the kite tail that will work best for this type of kite. The kite design is basic; it is made with a sheet of paper and a plastic straw. In this activity, students will investigate which point of attachment allows the kite to fly best. "Best" is operationally defined in terms of the way that the kite files:

a) "dances wildly" out of control or dives around in an unstable manner high in the sky with a loose string

b) "pulls and dives" aimlessly close to the ground

c) "floats smoothly," looking light and stable in the air

Letter "c" is clearly the most desirable kite performance. Tell students they must determine the best attachment point for the kite to fly smoothly: front point (6 cm), midpoint (7 cm), or rear point (8 cm).

Whether or not students have made and/or flown kites, they probably have seen kites flying. Discuss what a kite looks like when it is flying smoothly, high in the sky. Let students know they will be investigating variables that affect the way a kite flies.

Have students make the straw kite (see kite directions on pages 150–152). When the kites are completed, ask students to determine which of the three attachment points will make the kite fly the best.

Students may conclude that, in general, the kite did not fly well at any attachment point. Explain to students that the kite will fly beautifully with the right tail and attachment point; they will have a chance to change kite tails in activity 2.

# Kite Directions

To make a kite, you need one sheet of 8.5 x 11 inch paper, one plastic straw, and tape.

1. Fold the sheet of paper in half the long way. Each half sheet of paper will be 5.5 inches wide.

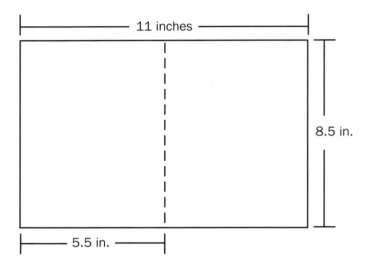

2. Measure 3 cm from the folded edge and draw a line from top to bottom.

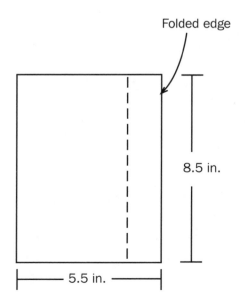

3. Crease the paper at the line and fold it over. The folded section is called the keel of the kite. Place the keel face down and open the paper, leaving the folded part (keel) underneath. Tape along the center line.

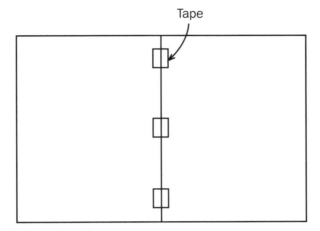

4. With the open paper facing you and the keel underneath, measure 2 cm from the top of the kite and draw a line across the kite. Tape a plastic straw along the line and secure it with a piece of tape at both ends and one in the middle. Turn the kite over and straighten out the keel.

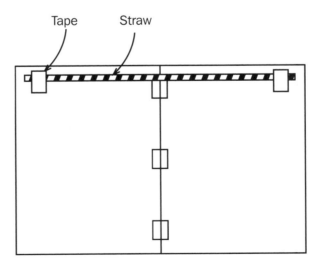

5. Measure 6 cm from the top of the kite. Put a mark on the keel about 1.5 cm from the edge at the 6 cm (from top) spot. Then measure another centimeter from the top and put another mark at 7 cm and another at 8 cm. Reinforce the marks with tape and punch a hole at the 6 cm, 7 cm, and 8 cm marks. These three holes will be the attachment points.

6 cm = front point

7 cm = midpoint

8 cm = rear point

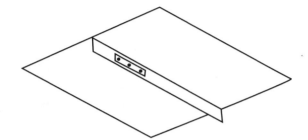

6. Make a tail for the kite from paper, plastic, tissue paper, etc. If all kites are to be the same, the group should decide on dimensions for the kite tail. If the tail is to be a part of the experiment, let the kite maker be creative.

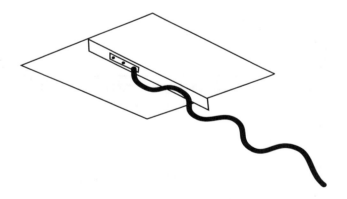

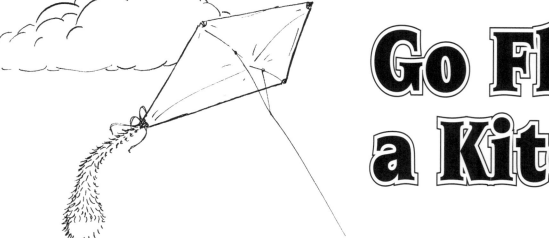

# Go Fly a Kite

Name: _____

Date: _____

You have been asked to help the local park district investigate kite aerodynamics. Park district officials have developed a new design for a kite that can be used by young children successfully, but they need help in determining where to attach the tail of the kite. Park officials know that the tail influences how smoothly a kite flies. The best tail attachment point is the one that helps the kite fly smoothly.

Make a kite using the directions provided and select a design, type of material, and length for the tail.

Which point of attachment will allow your kite to fly with a smooth, floating movement? Consider these three attachment points:

>   front point   (6 cm)
>
>   midpoint      (7 cm)
>
>   rear point    (8 cm)

Predict at which point you think the kite will fly best.

I predict _____

because _____

Determine a plan for answering the research question. Provide a step-by-step explanation of your plan in the space below.

 *IRI/Skylight Training and Publishing, Inc.*

Carry out your investigation. Be sure to keep accurate records of data collected.

**Description of the Kite's Tail**

Material used: _____

Length in cm: _____     Width in cm: _____

**Attachment Point**          **Description of Kite's Flight**

front point (6 cm) _____

midpoint (7 cm) _____

rear point (8 cm) _____


Based on what you learned through your experimentation, what can you conclude about the best attachment point for your kite?

_____

_____

_____

_____

_____

_____

_____

_____

_____

_____

PERFORMANCE
ACTIVITY 2

# Smooth Tailing

## Description of Activity

Now students will investigate different tail designs and/or lengths to determine what effect they have on the performance of the kite. Students will decide individually what tail designs and/or lengths they would like to compare to their first design. For each new design or length, the students will test the front, mid, and rear attachment points to determine, for the new design, which attachment point will give the best kite performance.

This performance assessment requires students to set up a minimum of two new experiments. For each experiment, students should communicate a clear description of their new design, document their testing procedures, and keep accurate records of the test results. The structure of activity 1 should prepare them to handle activity 2, since a similar experimental design can be used. Note that if students set up an experimental design that is different from the structure of activity 1, the data collecting and analysis will be more complex. This complexity results from students' changing both the variable and the testing procedure.

Because the kite tails are different for each student, experimental results should be different. Thus, students can go out and fly their kites, record their individual results, and keep journal entries relating to the types of tail materials they use, the lengths of the tails, the cause-and-effect relationships, environmental conditions, and data collected. Since there are so many possibilities for lengths of tails and types of materials, the students can report only on their own conclusions and determine the best of what has been tested by them. Hopefully, students will realize that the more types and lengths of tails that are tested, the more data they will have upon which to draw their conclusions.

## Presenting the Activity

Remind students that although their kites may have flown well in activity 1, there is always the possibility that they might fly better with different tails. Additional investigations testing different tail designs may provide the best overall design for the kite that the park district can use. Tails can be made of paper, plastic, tissue paper, etc., and they can vary in length.

As a dedicated researcher, students are asked to try at least two different tail designs besides the original one from activity 1 to determine if they can get their kites to fly better. They may change the material, the length, or both. For each tail design, students should test the three attachment points to determine for each new design which attachment point will give the best (smoothest) kite flight.

## MATERIALS

- ❏ simple paper and straw kite from activity 1
- ❏ kite string
- ❏ various types of paper strips such as newspaper, computer paper, tissue paper, etc.
- ❏ strips of plastic from garbage bags, bread bags, etc.
- ❏ scissors
- ❏ tape
- ❏ metric rulers or meter sticks

# Smooth Tailing

**In what ways can you change the tail of your kite to get a better flight performance?**

For each new design, test the three attachment points on the kite as you did in the first activity.

### New Tail Design I

Describe the new tail. Tell how it is different from the kite tail in activity 1.

New tail: _____

How it differs from the tail in activity 1: _____

_____

_____

Describe your plan for testing the new kite tail design.

_____

_____

_____

_____

_____

_____

Conduct the experiment and record observations and other data.

_____

_____

_____

Draw a conclusion based on the data you collected.

_____

_____

_____

### New Tail Design II

Describe the new tail. Tell how it is different from the kite tail in activity 1.

New tail: _____

How it differs from the tail in activity 1: _____

_____

Describe your plan for testing the new kite tail design.

_____

_____

_____

Conduct the experiment and record observations and other data.

_____

_____

_____

Draw a conclusion based on the data you collected.

_____

_____

_____

# Kite Design

In appreciation of your help in experimenting with factors that work best for the park district's model kite, the park district has given you a free pass to the pool for the summer.

A reporter from the local newspaper has asked you to write a feature article describing the processes and results of your experimentation with kite aerodynamics. She is interested in sharing the results of your work with the community, and the park district is anxious to get some publicity to promote the kite program.

Please help by writing an article that describes your research. Include a final conclusion after considering the data from your three experiments as well as the reports of other members of your class.

| **Indicators** | **Dimensions** |
|---|---|

### ACTIVITY 1 **Go Fly a Kite**

| | |
|---|---|
| Predicts based on logical reason | S |
| Gives a description of the tail of the kite | C |
| Provides measurements of tail in cm | S |
| Develops a plan for testing | S |

   (Teachers may determine what specific criteria they want to
see in the plan. Such criteria might include:

- identification of the independent and dependent variables
- strategy for experiment that is logical and reasonable
- steps described in sequence
- categories of data clearly described)

| | |
|---|---|
| Designs a data table | S |
| Records data for attachment points and flight performances | S |
| Draws a conclusion based on data | S |

### ACTIVITY 2 **Smooth Tailing**

| | |
|---|---|
| Describes the new tail design | C |
| Describes differences in design | S |
| Predicts based on logical reasoning | S |
| Describes plan | S |

  (again, additional criteria may be added)

| | |
|---|---|
| Designs a data table | S |
| Records data of flight performance for attachment points | S |
| Draws conclusion based on data | S |

**Suggested Criteria for Writing Prompt:** Article should include an accurate description of procedures and data. Dimensions of language arts may also be assessed, such as the mechanics of writing a paragraph.

# Role-ing through Science and Technology

## CHAPTER 13

## Targeted Process Skills

- applying concepts
- researching information
- communicating

## Rationale

Throughout their study of science topics, students should become increasingly aware of the vast number of occupations that are possible in the fields of science and technology. Students should read about occupations in science and technology and interview adults and professionals in the related fields whenever possible.

It is not unusual for aspiring architects and others in fine arts–related fields to develop portfolios of their work to take to job interviews. Besides observing physical and personality traits during an interview, a prospective employer can view the work of the candidate and get a first-hand impression of the artist's style and ability. Based on this model, students are asked to consider how other professional candidates "sell themselves" to prospective employers.

This performance assessment is probably best done at the end of a school year, since it allows students to apply career information they have collected throughout the year. The assessment task asks students to creatively design a performance for a job interview in order to show an understanding of concepts, skills, and habits of mind that are related to professions in science and technology.

Activity 1 is a basic research activity. Students are asked to find information related to a career in one of the fields of science they have studied. They will create a photo journal by selecting four significant features of their profession and drawing a picture that represents each feature and/or themselves as professionals involved in a significant aspect of the profession.

> ## PROJECT 2061 STATEMENTS
> ### from *Benchmarks for Science Literacy*
>
> People can learn about others from direct experience, from the mass communications media, and from listening to other people talk about their work and their lives (p. 154).
>
> • • •
>
> Each culture has distinctive patterns of behavior, usually practiced by most of the people who grow up in it (p. 155).
>
> • • •
>
> As students begin to think about their own possible occupations, they should be introduced to the range of careers that involve technology and science, including engineering, architecture, and industrial design. Through projects, readings, field trips, and interviews, students can begin to develop a sense of the great variety of occupations related to technology and to science and what preparation they require (p. 46).

In activity 2, students are given time to plan a presentation for a job interview that relates to a specific career area in science or technology. The assessment allows students to show that they understand the importance of the concepts, skills, and habits of mind that underlie science in a science- or technology-related profession. There is a powerful S-T-S connection here since students will be analyzing and describing the profession in terms of the "big picture" of science. A specific set of criteria should be identified to assess that the important dimensions are recognized by the student and are included in the performance. (The criteria may limit creativity somewhat, but they will be an asset in scoring the presentation and being accountable for a range of dimensions.)

You may wish to discuss with students what they would look for as an employer of one or more of the following: a babysitter for a young sibling, a housekeeper, or a coach. There should be some discussion of the need for a knowledge base, various job-related skills, and attitudes and values that are critical to good performance of the job. Certainly, one would expect that a babysitter enjoys small children, is responsible, and is safety conscious; one would expect a coach to be motivated, enthusiastic, organized, and under-standing. Other habits of mind related to science and technology (as identified in *Benchmarks for Science Literacy*), such as honesty, willingness to consider evi-dence, being a team player, and being logical and reasonable, might also be considered in the discussion.

# A Photo Journal of My Profession

PERFORMANCE ACTIVITY

## Description of Activity

This activity asks students to research a career related to one of the areas of science that they studied during the year. Students may use any number of resources, including human resources. Their information should include concept understanding, skills, and attitudes and values needed to be successful in the profession. Information related to the amount of education and/or training needed for entry level in the profession would also be helpful.

## Presenting the Activity

Ask students to think about some profession with which they are familiar. Perhaps they can discuss the profession of a parent or relative. Teachers can engage students in a discussion of the teaching profession, if necessary.

Tell students to consider the following factors relating to a profession—the knowledge base, the skills needed in the profession, and the values and attitudes necessary for success in the field. For example, for a profession that deals with living animals, one should have a love of animals and an overall respect for living things.

Students will research a profession related to an area of science or technology they studied this year (or to other areas of science, if you wish). They should develop a list of questions to research about the profession they choose to investigate, such as, What sort of education and/or training is necessary for this job? Where can one get the necessary education? What sort of knowledge base does the professional need? What kinds of skills does the professional need for his or her work? Are there any certifications or licenses required? What kinds of equipment are used in the work? Does the use of the equipment require skill and/or training? What is society's attitude toward this profession? What attitudes or values would one need to be successful in this profession? Of what value is the profession to society? Approximately what does the job pay per year?

In this activity, students will write a report with the information they discovered. Then, they will select four significant features of the profession that they can draw and describe. Each of these features will be represented by a "photograph" of some aspect of the profession or of themselves (as the professional) involved in the profession. They should draw each picture in the camera lens provided and describe the significant feature of the photo. The four pictures and descriptions represent a photo journal of the profession.

### MATERIALS

❑ access to library materials, human resources, computers, databases, etc. for career information

# A Photo Journal of My Profession

Name: _____

Date: _____

**Photo I**

The profession I am presenting is _____

The science topic it is most closely related to is _____

Title: _____

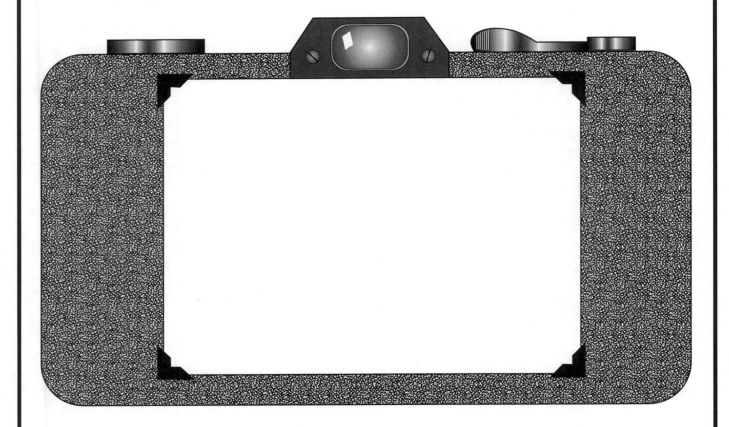

Description of photo:

_____

_____

_____

_____

_____

# A Photo Journal of My Profession

Name: _____

Date: _____

**Photo II**

The profession I am presenting is _____

The science topic it is most closely related to is _____

Title: _____

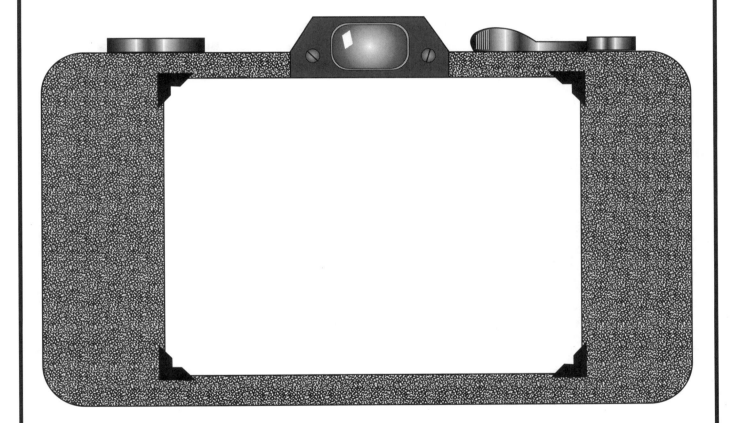

Description of photo:

_____

_____

_____

_____

_____

# A Photo Journal of My Profession

Name: _____

Date: _____

**Photo III**

The profession I am presenting is _____

The science topic it is most closely related to is _____

Title: _____

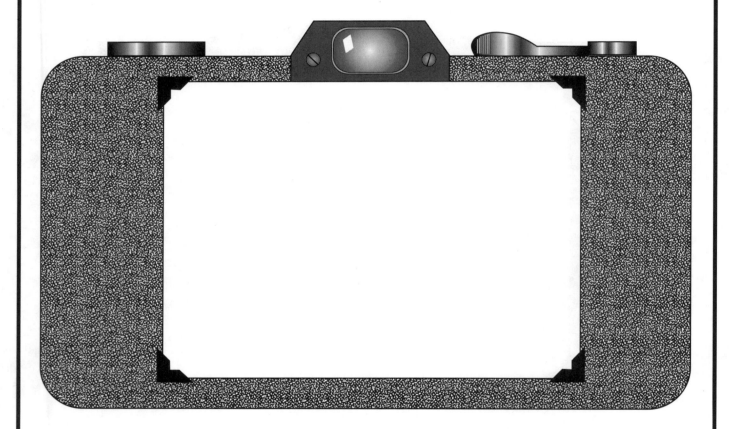

Description of photo:

_____

_____

_____

_____

_____

*IRI/Skylight Training and Publishing, Inc.*

# A Photo Journal of My Profession

Name: _____

Date: _____

**Photo IV**

The profession I am presenting is _____

The science topic it is most closely related to is _____

Title: _____

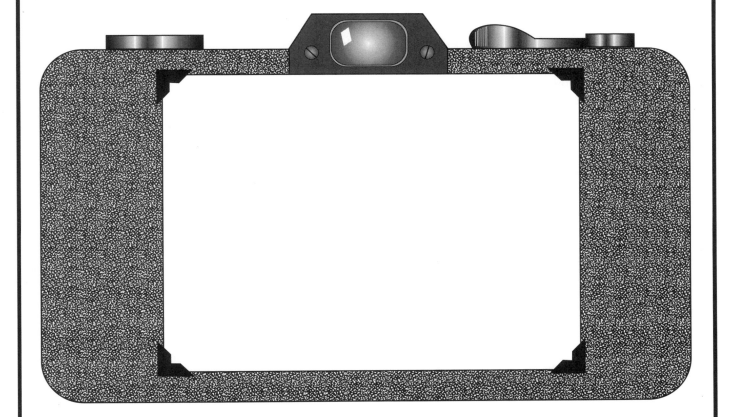

Description of photo:

_____

_____

_____

_____

_____

## PERFORMANCE ACTIVITY 2

# Designing and Performing a Creative Interview, or "I Really Need This Job"

### Description of Activity

Students will design a portfolio, plan a multimedia presentation, compose a song or rap, choreograph a dance, or design some other creative way to interview for a job in the profession they have selected to research.

### Presenting the Activity

Tell students that competition for professional positions is high, so they will develop and present a unique way to apply for a job. Their presentations may consist of such things as a portfolio, a multimedia presentation, a song or a rap, or a dance, and should include evidence that they have a grasp of or ability to demonstrate the following:

- the necessary concept knowledge base for the job
- skills that are needed for the profession
- attitudes, values, and other habits of mind that one would need to be successful in the field
- an understanding of the importance of the profession to society

There are no student response pages for this activity. The product will be a portfolio, a videotape of a presentation, or an actual presentation (song, rap, dance) that may include props, posters, mobiles, etc. It is important that the presentation be convincing, since it is an interview for a position. Students should realize that there may be stiff competition in their field of interest.

### MATERIALS

- ❏ AV equipment (camcorder, VCR, audiotape or CD player, overhead projector, etc.)
- ❏ audiotapes or CDs
- ❏ blank videotapes
- ❏ poster board
- ❏ overhead transparencies
- ❏ markers
- ❏ other materials as needed

**Indicators**                                                    **Dimensions**

## ACTIVITY 1  A Photo Journal of My Profession

Completes a research report about a profession related to
    science and/or technology                                    C, STS, H
Uses a minimum of two sources of information (optional)            C
Includes information relating to all of the research questions
    determined by the class (may give one point to each question)  C, STS, H
Provides the name of the profession and the area of science
    it relates to                                                 C, STS
Selects four significant features of the profession               C
Completes information for each photo entry
• picture of significant feature                                  C, STS
• description of significant feature                              C, STS

## ACTIVITY 2  Designing and Performing a Creative Interview

Shows a concept knowledge base for the profession                 C
    (This may be further stated as, describes a minimum of four
    concept-related areas the professional would need to have.
    For example, a paleontologist would need to know the various
    types of fossils and how they were formed.)
Demonstrates skills and/or understanding of skills                S
    (This may be further stated as, demonstrates at least two skills
    the professional would need and describes at least three others.
    For example, a microbiologist would need to have excellent
    skill using a microscope.)
Displays understanding of attitudes, values, and/or habits of mind  H
    (This might include a variety of qualities related to job
    responsibility and the profession: inquisitiveness, responsibility,
    open-mindedness, etc.)
Describes importance of the profession to society                 STS

This rubric might assess other aspects of the performance that are more subjective. Some of these criteria might include creative, appealing, convincing, able to put all the "pieces" together, etc. For each criterion, a description of what is acceptable and what is unacceptable is necessary. Students may be involved in determining how these factors will be scored.

# Bibliography

Aldis, R. 1991. *Rain Forests*. New York: Dillon Press.

Alverno College Faculty. 1985. *Assessment at Alverno College*. Rev. ed. Milwaukee: Alverno College.

American Association for the Advancement of Science (AAAS). 1989. *Science for all Americans*. New York: Oxford University Press.

————. 1993. *Benchmarks for science literacy*. New York: Oxford University Press.

American Council on the Teaching of Foreign Languages. 1982. ACTFL *provisional proficiency guidelines*. Hastings-on-Hudson, N.Y.: ACTFL Materials Center.

American Psychological Association. 1985. *Standards for educational and psychological testing*. Washington, D.C.: American Psychological Association.

Anderson, S. R. 1993. Trouble with testing. *The American School Board Journal* 180 (June): 24–26.

Archbald, D., and F. Newmann. 1988. *Beyond standardized testing: Authentic academic achievement in the secondary school*. Reston, Va.: NASSP Publications.

Astin, A. W. 1991. *Assessment for excellence: The philosophy and practice of assessment and evaluation in higher education*. New York: American Council on Education, Macmillan.

Baker, E. L. 1993. Questioning the technical quality of performance assessment. *The School Administrator* 50 (December): 12–16.

Baker, E. L., et al. 1991. *Cognitively sensitive assessments of student writing in the content areas*. Los Angeles: The National Center for Research on Evaluation, Standards and Student Testing (CRESST).

Baker, E. L., H. F. O'Neil, Jr., and R. L. Linn. 1993. Policy and validity prospects for performance-based assessment. *American Psychologist* 48 (12): 1210–18.

Belanoff, P., and P. Elbow. 1986. Using portfolios to increase collaboration and community in a writing program. *Writing Program Administrator* 9 (1): 27–40.

Berk, R. A., ed. 1986. *Performance assessment methods and applications*. Baltimore: Johns Hopkins University Press.

Berlak, H., et al. 1992. *Toward a new science of educational testing and assessment*. New York: State University of New York Press.

Bloom, B. S., ed. 1956. *Taxonomy of educational objectives. Book 1: Cognitive domain*. New York: Longman.

Bloom, B. S., G. F. Madaus, and J. T. Hastings. 1981. *Evaluation to improve learning*. New York: McGraw-Hill.

Bond, L., L. Friedman, and A. van der Ploeg. 1993. *Surveying the landscape of state educational assessment programs*. Washington, D.C.: Council for Educational Development and Research and The National Education Association.

Bracey, G. W. 1993. Testing the tests. *The School Administrator* 50 (December): 8–11.

Brandt, R., ed. 1992. *Performance assessment: Readings from educational leadership*. Alexandria, Va.: Association for Supervision and Curriculum Development.

Burke, K. 1994. *The mindful school: How to assess authentic learning*. Palatine, Ill.: IRI/Skylight Publishing and Training.

California Assessment Program. 1989. *Guidelines for the mathematics portfolio: Phase II pilot working paper*. Sacramento: California Assessment Program (CAP) Office, California State Department of Education.

California State Department of Education. 1989a. *A question of thinking: A first look at students' performance on open-ended questions in mathematics*. Sacramento: California State Department of Education.

————. 1989b. *Writing achievement of California eighth graders: Year two*. Sacramento: California State Department of Education.

Center on Learning, Assessment, and School Structure. 1993. *Standards, not standardization. Vol. III, rethinking student assessment*. Geneseo, N.Y.: Center on Learning, Assessment, and School Structure.

Charles, R., and E. Silver, eds. 1988. *The teaching and assessing of mathematical problem solving. Vol. 3*. Reston, Va.: National Council of Teachers of Mathematics.

## Bibliography

College Board. 1986a. Evaluating the AP portfolio in studio art. In *Advanced Placement in Art*. Princeton, N.J.: Educational Testing Service, CEEB.

————. 1986b. General portfolio guidelines. In *Advanced Placement in Art*. Princeton, N.J.: Educational Testing Service, CEEB.

Connecticut Department of Education. 1990. *Toward a new generation of student outcome measures: Connecticut's common core of learning assessment*. Hartford, Conn.: State Department of Education, Research and Evaluation Division.

Cox, D. 1987. Concept-based science instruction. Presentation, Portland State University, Portland, Oreg.

Cronbach, L. J. 1989. *Essentials of Psychological Testing*. 5th ed. New York: Harper & Row.

Department of Education, New Zealand. 1989. *Assessment for better learning: A public discussion document*. Wellington, New Zealand: Department of Education.

Department of Education and Science and the Welsh Office (U.K.). 1988. *National curriculum: Task group on assessment and testing: A report*. London: Department of Education and Science and the Welsh Office.

————. 1989. *English for ages 5 to 16: Proposals of the secretary of state for education and science*. London: Department of Education and Science and the Welsh Office.

Educational Testing Service. 1986. *The redesign of testing for the 21st century: Proceedings of the ETS 1985 invitational conference*. Princeton, N.J.: Educational Testing Service.

————. 1993a. *Linking assessment with reform: Technologies that support conversations about student work*. Princeton, N.J.: Educational Testing Service.

————. 1993b. *What we can learn from performance assessment for the professions: Proceedings of the ETS 1992 invitational conference*. Princeton, N.J.: Educational Testing Service.

Edwards, J. 1991. To teach responsibility, bring back the Dalton Plan. *Phi Delta Kappan* 72 (January): 398–401.

Elbow, P. 1981. *Writing with power: Techniques for mastering the writing process*. New York: Oxford University Press.

————. 1986. *Embracing contraries: Explorations in learning and teaching*. New York: Oxford University Press.

Elwell, P. 1991. To capture the ineffable: New forms of assessment in higher education. In *Review of research in education*, edited by G. Grand. Washington, D.C.: American Educational Research Association.

Falk, B., and L. Darling-Hammond. 1993. *The primary language record at P. S. 261: How assessment transforms teaching and learning*. New York: NCREST, Columbia University.

Feuer, M. J., K. Fulton, and P. Morrison. 1993. Better tests and testing practices: Options for policy makers. *Phi Delta Kappan* 74 (March): 532.

Finch, F. L., ed. 1991. *Educational performance assessment*. Chicago: Riverside Publishing Company, Houghton Mifflin.

Fitzpatrick, R., and E. J. Morrison. 1991. Performance and product evaluation. In *Educational Measurement*, edited by R. L. Thorndike. 2d ed. New York: American Council on Education, Macmillan.

Fox, R. F. 1993. Do our assessments pass the test? *The School Administrator* 50 (December): 6–11.

Frederiksen, J. R., and A. Collins. 1989. A systems approach to educational testing. *Educational Researcher* 18 (December): 27–32.

Frederiksen, N. 1984. The real test bias. *American Psychologist* 39 (March): 193–202.

Gagnon, P., and The Bradley Commission on History in the Schools. 1989. *Historical literacy: The case for history in American education*. Boston: Houghton Mifflin.

Gardner, H. 1989. Assessment in context: The alternative to standardized testing. In *Changing assessments: Alternative views of aptitude, achievement, and instruction*, edited by B. Gifford. Boston: Kluwer Academic Press.

Gentile, C. 1991. *Exploring new methods for collecting students' school-based writing*. Washington, D.C.: United States Department of Education.

Glaser, R. 1971. A criterion-referenced test. In *Criterion-referenced measurement: An introduction*, edited by J. Popham. Englewood Cliffs, N.J.: Educational Technology Publications.

Grant, G. 1979. On competence: A critical analysis of competence-based reforms in higher education. San Francisco: Jossey-Bass.

Haney, W. 1985. Making testing more educational. Educational Leadership 43 (October): 2.

Haney, W., and L. Scott. 1987. Talking with children about tests: An exploratory study of test item ambiguity. In Cognitive and linguistic analyses of test performance, edited by K. O. Freedle and R. P. Duran. Norwood, N.J.: Ablex.

Hanson, F. A. 1993. Testing testing: Social consequences of the examined life. Los Angeles: University of California Press.

Hart, D. 1994. Authentic assessment: A handbook for educators. Menlo Park, Calif.: Addison-Wesley.

Herman, J., P. Aschbacher, and L. Winters. 1992. A practical guide to alternative assessment. Alexandria, Va.: Association for Supervision and Curriculum Development.

International Baccalaureate Examination Office. 1991. Extended essay guidelines. Wales, U.K.: International Baccalaureate Examination Office.

Jorgensen, M. 1993. The promise of alternative assessment. The School Administrator 50 (December): 17–23.

Kendall, J. S., and R. J. Marzano. 1994. The systematic identification and articulation of content standards and benchmarks. Update. Aurora, Colo.: Mid-continent Regional Educational Laboratory.

Kentucky General Assembly. 1990. Kentucky Education Reform Act (KERA). House Bill 940. Lexington.

Lamme, L. L., and C. Hysmith. 1991. One school's adventure into portfolio assessment. Language Arts 68 (December): 629–40.

Linn, R. 1986. Barriers to new test designs. In The redesign of testing for the 21st century: Proceedings of the 1985 ETS invitational conference. Princeton, N.J.: Educational Testing Service.

———. 1989. Educational measurement. 3d ed. American Council on Education Series on Higher Education. Washington, D.C.: American Council on Education.

———. 1993. Educational assessment: Expanded expectations and challenges. Educational Evaluation and Policy Analysis, 15 (1): 1–16.

Linn, R., E. Baker, and S. Dunbar. 1991. Complex, performance-based assessment: Expectations and validation criteria. Educational Researcher 20 (November): 15–21.

Lowell, A. L. 1926. The art of examination. Atlantic Monthly 137 (January): 58–66.

Maeroff, G. 1991. Assessing alternative assessment. Phi Delta Kappan 17 (December): 272–81.

Marzano, R. J., D. Pikering, and J. McTighe. 1994. Assessing student outcomes: Performance assessment using the dimensions of learning model. Alexandria, Va.: Association for Supervision and Curriculum Development.

McClelland, D. 1973. Testing for competence rather than for "intelligence." American Psychologist 28 (January): 1–14.

McCloskey, M., A. Carramazza, and B. Green. 1989. Curvilinear motion in the absence of external forces. Science 210: 1139–41.

Mehrens, W. 1992. Using performance assessment for accountability purposes. Educational Measurement: Issues and Practice 11 (spring): 3–9.

Messick, S. 1989a. Meaning and values in test validation: The science and ethics of assessment. Educational Researcher 18 (2): 5.

———. 1989b. Validity. In Educational measurement. 3d ed. American Council on Education Series on Higher Education. Washington, D.C.: American Council on Education.

Mills, R. 1989. Portfolios capture rich array of student performance. The School Administrator 11 (December): 46.

Ministry of Education, Victoria, Australia. 1990. Literacy profiles handbook: Assessing and reporting literacy development. Victoria: Education Shop; distributed in U.S. by TASA, Brewster, N.Y.

———. 1991. English profiles handbook: Assessing and reporting students' progress in English. Victoria: Education Shop; distributed in U.S. by TASA, Brewster, N.Y.

Mitchell, R. 1992. Testing for learning. New York: Free Press, Macmillan.

Moss, P. A. 1994. Can there be validity without reliability? Educational Researcher 23 (March): 5–12.